DISCOVERING DINOSAURS

FRONTISPIECE **Although not a paleontologist, Roy Chapman Andrews was the intrepid leader of the American Museum of Natural History's Central Asiatic Expeditions. These expeditions put Asia on the map as one of the greatest fossil-bearing regions on the planet.**

DISCOVERING DINOSAURS

EVOLUTION, EXTINCTION, AND THE LESSONS OF PREHISTORY

MARK NORELL   LOWELL DINGUS   EUGENE GAFFNEY

A PETER N. NEVRAUMONT BOOK

UNIVERSITY OF CALIFORNIA PRESS

BERKELEY LOS ANGELES LONDON

University of California Press
Berkeley and Los Angeles, California

University of California Press, Ltd.
London, England

First Paperback Printing 2000

Library of Congress Cataloging-in-Publication Data

Norell, Mark.
Discovering dinosaurs: evolution, extinction, and the lesson of prehistory / Mark Norell, Eugene Gaffney, Lowell Dingus.
p. cm.
Rev. ed. of: Discovering dinosaurs in the American Museum of Natural History: New York: Knopf, 1995.
Includes index.
ISBN 0-520-22501-5
1. Dinosaurs. I. Gaffney, Eugene S. II. Dingus, Lowell. III. Norell, Mark. Discovering dinosaurs in the American Museum of Natural History. IV. Title.
QE861.4.N67 2000
567.9—dc21 99-053335

Printed in Italy by Editoriale Bortolazzi-Stei
Updated & Revised Edition

This book was created and produced by
Nevraumont Publishing Company, Inc.
New York, New York
President: Ann J. Perrini

Book Design: Barbara Balch
Jacket Design: Frances White

9 8 7 6 5 4 3 2 1

The paper used in this publication meets the minimum requirements of American National Standard for Information Sciences—Permanence of Paper for Printed Library Materials, ANSI Z39.48-1984

CONTENTS

II THE GREAT HALL OF DINOSAURS *97*

INTRODUCTION

One of the intriguing properties of life is its overwhelming diversity, and one of the most important pursuits of science is to study this diversity by comparing and contrasting the characteristics of different organisms. This study, called systematics, is the basis for reconstructing evolutionary history.

During the past 30 years, a more rigorous approach to systematics has been developed. Called cladistics, this method organizes animals into groups based on uniquely evolved characteristics that they share. Cladistics arranges animals into groups called clades, which contain the first member of the group and all of its descendants. Although this is a highly technical and far from problem free enterprise, the basic approach is simple.

We see a pattern in the diversity of life when we look for characteristics shared by different organisms. Using this pattern of characteristics, we can arrange organisms into smaller groups within larger groups. The arrangement of groups within groups results from organisms evolving when descendants inherit modified characteristics from their ancestors. By studying the distribution of these modified characteristics in different animals, we can determine the order in which they evolved and thereby infer the sequence of evolutionary history.

Branching diagrams called cladograms show the sequence of evolution. Beginning at the lowest branching point and reading up branch by branch, we can reconstruct the order in which new modified characters—or, as they are known in cladisitcs, "derived characters"—evolved. Each newly derived character distinguishes a clade that contains all the descendants on higher succeeding branches. The sequence of derived characteristics thus defines a pattern of groups within groups.

For example, many animals share some characters. Bony fishes, frogs, lizards, and humans all have a backbone composed of small individual bones called vertebrae. Thus, they are all included in the group, or clade, that we call vertebrates. We infer that all vertebrates arose from the first animals that had a backbone, and this first vertebrate is called the common ancestor of all vertebrates. Fewer animals within the vertebrate clade share other characteristics. For instance, frogs, lizards, and humans (but not fish) have four limbs with bony wrists, ankles, fingers, and toes. These animals are called tetrapods ("four-footed").

Using cladistics as the system for studying dinosaurs has important ramifications. Because the system is based on evidence rather than on subjective belief, our theories (that is, anything we propose about dinosaurs—how they are related to each other and other animals, their color or how fast they could run) must be based on available evidence, not on speculation, preconceived ideas, or authority.

These theories are continually tested as new evidence is gathered. If the new evidence conforms to the predictions of the theory, it passes the test (although it can never be shown to be absolutely true). If the prevailing theory does not explain the new observation, then the theory is rejected and replaced with a new one that that is more consistent with the evidence. A limitation of this process is that surprisingly large numbers of questions can not be examined critically because not enough data that focuses directly on the problem can be collected. Investigators must step back, adopt a cautious attitude, and recognize that these problems are not easily resolved. In this book we use the basic reference system of cladistics to examine dinosaurs.

We ended the first edition of this book with a statement that "We hope that these specimens will provide us with important information that will render much in this book inaccurate." When we wrote that five years ago, even we did not expect the onslaught of discovery that would close the last years of the century. Important specimens have been found on every continent and several islands. In addition, new technologies have come into play to locate dinosaur fossils, to prepare them and to analyze their structure and relationships to other dinosaurs and animals. We will review some of these advances before going on to "Discover Dinosaurs."

FIGURE 1. Modern paleontologists use pools both traditional and modern. While there is no substitute for standing on the top of a car with a pair of fine binocular, electronic GPS units allow one to place themselves within meters of where you are.

ADVANCES IN COLLECTION

Increasingly electronic tools are being used to hunt dinosaurs. A few expeditions have begun to deploy highly experimental devices, allowing paleontologists to "see" through rock to locate specimens. Usually this is accomplished by setting off small vibrations or explosions. The sound then travels through the sediment at different rates depending on the densities of the sediment. Because fossils are usually a different density than the surrounding sediments, the sound waves travel at different rates through them. By using an array of small microphones to capture these sounds, computers reconstruct the three-dimensional structure below the surface and actually "see" the subsurface bones on a computer screen. This technology is only in its infancy. It has been used, however, at some fossil sites; for example, the excavation of the large sauropod Seismosaurus in New Mexico. A number of the

Seismosaurus' bones were located several feet beneath the surface using this technology. It allowed the excavators to develop informed strategies for removing the fossil of this giant animal from the hard rock before they commenced digging.

Of much greater utility is the use of satellites. Satellites are used for two reasons: positioning and imaging. Making your way through the wilderness, especially when you are unfamiliar with the landmarks or even traversing well known country at night or in bad weather can make for tough going. Locating known sites may also be hard. Some of the locality information is sketchy, like one entry in our catalogue; "approximately 125 miles North of Miles City." Today, with a simple and inexpensive instrument called a Global Positioning System (GPS), it is possible to locate yourself within a few feet anywhere on the planet. Now there is no excuse for getting lost (unless the map that you are plotting your position on is inaccurate—which sometimes happens!) or "losing" a previously visited locality.

While we are still a long way from being able to locate single dinosaur specimens from space images, pictures taken from satellites allow you to choose routes through difficult terrain and give you a 'feel' for an area before you actually arrive. They are even used as exploration tools. In the Gobi Desert, the focus of the recent Mongolian Academy of Sciences-American Museum expeditions has been a series of bright-red and orange sandstones. Fossils of dinosaurs and other animals contemporary to them have previously been found in similar rock formations.

With digital satellite images it is possible to isolate certain spectral color signals. If we know we are looking for a particular color of rock, these particular spectral bands can be boosted and enhanced. The end product is a kind of treasure map taken from space, where possible fossil localities (even areas as small as 10 x 10 meters) jump out at you on the map. Use of satellite imagery and GPS has contributed to the success of the Gobi expeditions and others around the globe.

ADVANCES IN PREPARATION

Preparation is still one of the most important parts of getting a specimen out of the ground and onto a researchers desk. Preparators are refining their trade constantly. These advances include developing new techniques to remove specimens from their encasing matrix, experimenting with the latest in new materials to infuse or harden bones that would naturally break apart once removed from their supportive rock matrix, and finding creative ways to keep fossils preserved for decades and longer in museum collections.

Electronic technology is beginning to have an impact on preparation as well. Making a reproduction of a fragile specimen is very difficult. Traditional casting requires that a mold be made using silicon rubber or something similar. This places a fragile specimen in great jeopardy. It is now possible to use digital information taken from CAT scans—or more typically, a laser surface scanner—to develop exact copies of fossil bones. In this procedure the laser samples thousands of points on the surface of a bone, giving each point a coordinate in three dimensions. The bone is unharmed because nothing physically touches its surface. The points are then put into a computer, and a surface is drawn connecting them. This is the same tech-

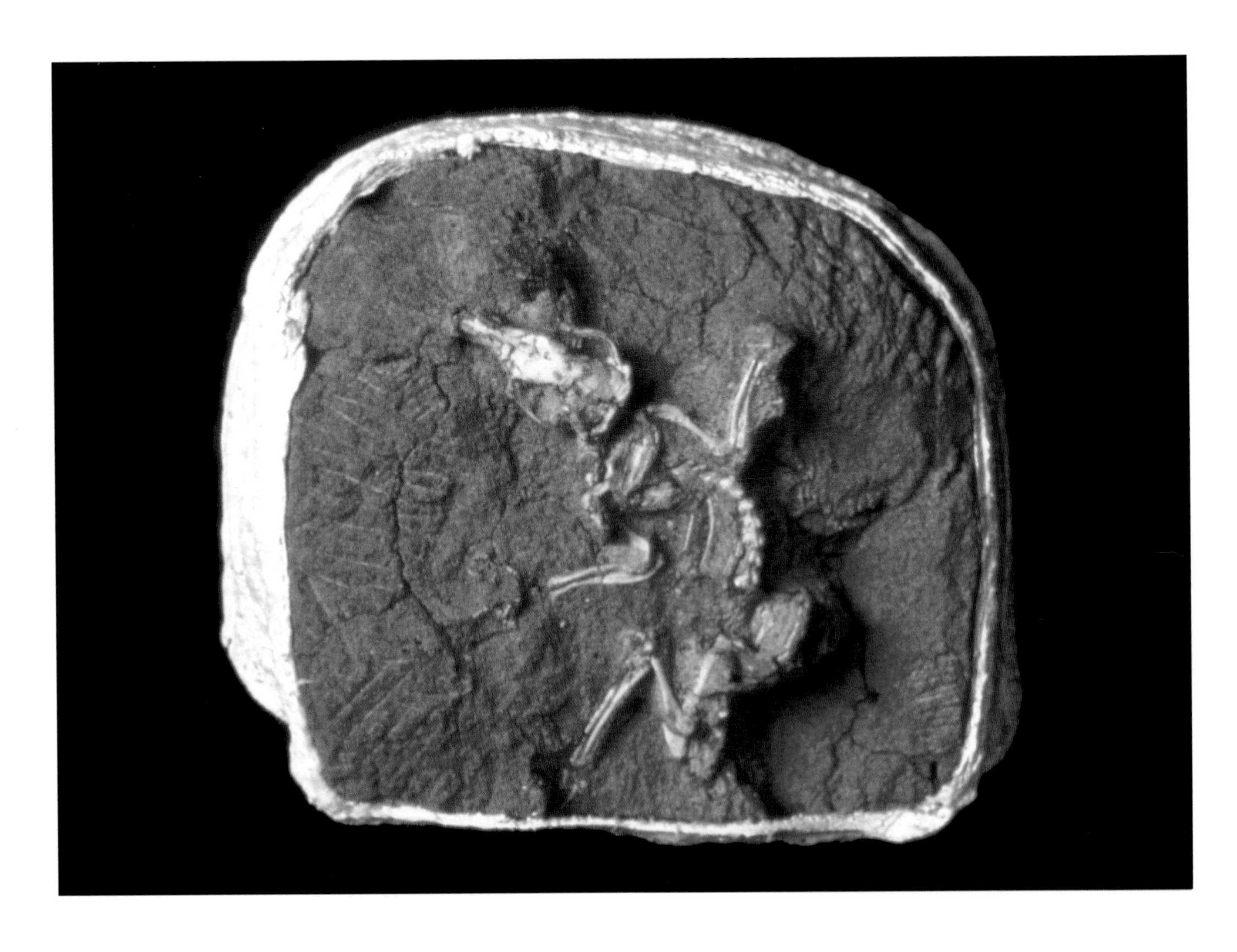

FIGURE 2. Ukhaatherim is a primitive mammal from the locality of Ukhaa Tolgod. It is an advanced animal that gave birth to live young. Skeletons such as these are the finest representatives of Mesozoic mam-mals collected anywhere in the world.

nology, and in fact much of the same software, used to make special effects for movies like The Terminator and Star Wars. Using special machines, it is even possible to construct models using this digital information. In years to come, as computers and networks become faster, it will be possible to send digital scans to researchers across the planet for them to instantly construct three-dimensional models of skeletal elements.

ADVANCES IN ANALYSIS

Use of digital imaging combined with CAT scanning has given us a powerful new tool to analyze specimens. Some specimens do not even need conventional preparation, as it is now possible to see in and through them without taking them out of their matrix. This technology is not a substitute for preparation, as traditional techniques are still required to see the majority of features. Instead it is an adjunct that allows all kinds of new character systems to be studied. As an example, certain theropod dinosaurs (including birds, as you will read later,) have a complex system of air passages in their skull bones. Conventionally, the only way to study these skull bones is to break them—hardly a legitimate practice on a one-of-a-kind fossil specimen. CAT scanning allows us to examine these "un-preparable" features for the first time and has provided access to new, entirely un-sampled character systems.

CAT scanning and digital laser scanning are also useful in reconstructing the skeletons of organisms. This is usually done to analyze skeletal structure or function. Skeletons can be scanned and then animated to move as if they were living animals. Skeletons, however, are rarely complete. If parts of the skeleton can be digitally captured (either through laser scanning or CAT), corresponding parts can be constructed. Similarly, composite animals can be constructed from the bones of two or more animals of the same species. Even if the bones are different sizes, it is simple to scale them to equal size.

The main goal of many dinosaur paleontologists is to discover the relationships among dinosaurs. Using these relationships we can then begin to make sense about all the other things we discover. This is the cladistic analysis discussed earlier. As we pointed out, it is the foundation for understanding the evolution of characters like behavior (*see* "The Nesting Oviraptor," page 191) and feathers (see "Caudipteryx," page 188).

Until recently, doing cladistic analysis on large data sets of fossil animals was hampered by a number of technical difficulties. We lacked ideas about how to deal with missing data; combining data of disparate types was also difficult. And our methods for searching for the family trees that best explained the evidence were imprecise when dealing with more than just a few species.

Progress has been made. Largely, this has come through the development of new methodological tricks, as well as greater computing power. Currently, systematists employ some of the fastest computers in the world for analyzing large and complex data sets.

ADVANCES IN DISCOVERY

Interesting new animals have been found in the Sahara, South America, Madagascar, Asia and Europe.

After nearly 100 years as the king of carnivorous dinosaurs, Tyrannosaurus rex has been dethroned. Discovery of giant theropods, like the Early Cretaceous Carcharodontosaurus in Morocco and the approximately 95 million year old Giganotosaurus in Africa shows clearly that giant theropods, even more giant than Tyrannosaurus, were not limited to the latest Cretaceous. Fantastic specimens of other Southern Hemisphere theropods have been discovered—including a wonderful skull of the theropod Majungathulus from Madagascar. In total these discoveries have shown that there is a group of closely related South American carnivorous dinosaurs, the Abelosauroidae, whose diversity and extent were totally unknown before these discoveries.

The largest dinosaur is now considered to be Argentinosaurus. This specimen is so large that just collecting all of it will take years. A single vertebrae (a backbone segment) stands in excess of 4 feet tall. This animal is at least 15% larger than the previous record holder. Finding land animals this large has important implications for studies of dinosaur physiology, behavior, and development. As the variance of possible sizes increases, challenges are presented to scientists trying to how these animals got so big, and how they could operate once reaching adult size.

EGGS AND BABIES

Eggs and embryos of dinosaurs are extremely rare. Recently, remarkable discoveries have come out of China, South America, and Mongolia. Not only have nests been found—some containing embryos—but adult dinosaurs sitting on top of nests have been recovered from Asia (*see* "The Nesting Oviraptor," page 191) and North America. Some of the embryos recovered from South America and China even have preserved soft tissue. One notable discovery (*see* "Auca Mahuevo," page 193) preserves sauropod embryos. These embryos are remarkable because the skin texture of the animals is preserved—the first clues we have about what the skin of some of the largest dinosaurs looked like.

Even though entire dinosaur eggs are rare, eggshell is very common at many localities. Dinosaur eggshell has a complex structure that is highly variable. Sorting this into an evolutionary view, or even identifying which eggs went with which dinosaurs, has been problematic up to now. The situation has improved because a significant number of dinosaur embryos have recently been collected. It is now possible to definitively tie specific types of dinosaurs to specific types of eggs. In the future this will allow us to do three new things: 1) use dinosaur eggshell as character evidence for refining our hypotheses of how dinosaurs are interrelated, 2) to determine which dinosaur lived at a certain time and place even if all we find are eggshells, and 3) to use dinosaur eggshell and eggs in behavioral and perhaps physiological studies.

In the earlier edition of this book we were clear in our support for the hypothesis that birds are living dinosaurs—that pigeons and chickens are more closely related to Tyrannosaurus rex than Tyrannosaurus rex is to dinosaurs as familiar as Stegosaurus and Triceratops. Discovery over the last five years has bolstered this view (*see* "Why are birds a kind of Dinosaur," page 11).

A number of new animals, many not yet described, have been found. Several of these impact the theory of the origin of birds. Some are animals that lie very close to the origin of the clade that contains birds: called Avialae. Animals such as Rahoonavis from Madagascar and Unenlagia from Argentine Patagonia are further blurring the distinction between the near ancestors of today's birds and what has been typically called a dinosaur.

Most notable among these are the spectacular specimens found in the Liaoning (*see* "Liaoning," page 214) area of China, about 320 Kilometers north of Beijing. Besides a diversity of early and primitive birds, these early Cretaceous beds have produced a wealth of small theropod dinosaurs. What is so spectacular about these fossils is that they preserve soft tissue. This soft tissue includes feathers and feather like structures. At first controversial, it now appears that animals showing an entire continuum documenting the stages of feather evolution have been discovered.

The most primitive of these animals is the small long-tailed Sinosauropteryx. Preliminary analyses indicate that it is a generalized theropod

FIGURE 3. Fieldwork in the Gobi is not all heat and sand. Continental climatic conditions are extreme from snow and hail to 120-degree heat and fierce wind.

dinosaur, not intimately related to Avialae (birds). It has a body covered with small filamentous fluff. These "protofeathers" are not branched like bird feathers and appear to be uniformly distributed on the body. Other more advanced animals like Caudipteryx and Protarchaeopteryx have true feathers—feathers with a median shaft and near perpendicular branches called barbs. Different sorts of feathers are found on different parts of the body in these animals with long feathers found on the arms and tails. The arm feathers are notable in that they lack the aerodynamic specializations found in flying birds. The bone structure of these animals clearly indicates that, while closely related, they are not part of the group that is commonly called birds. This means that feathers predate the origin of birds. Try this one on your imagination—we predict that animals like Velociraptor and Oviraptor had a feathered body covering. We predict that even Tyrannosaurus would have had a fluffy body covering at some stage of its life cycle. The last five years have certainly changed our conception of how we think of these animals.

Dinosaur science is rapidly evolving, and there are fascinating things yet to be learned. As we pointed out earlier, part of this is discovery, but a much more important part is the development of interesting and answerable questions. We hope that

THE EVOLUTIONARY RELATIONSHIPS OF DINOSAURS

AS DETERMINED BY CLADISTICS

Despite the popular interest of dinosaurs and the many new theories of their behavior, extinction, and ecology that appear frequently in the popular press, relatively few rigorous studies of their evolutionary history have been undertaken, probably because most people who study dinosaurs are interested in the more speculative aspects of dinosaur biology. This approach is unfortunate because understanding behavior, ecology, and other biological attributes depends on a firm foundation of evolutionary relationships. Although rigorous cladistics analyses are lacking for many

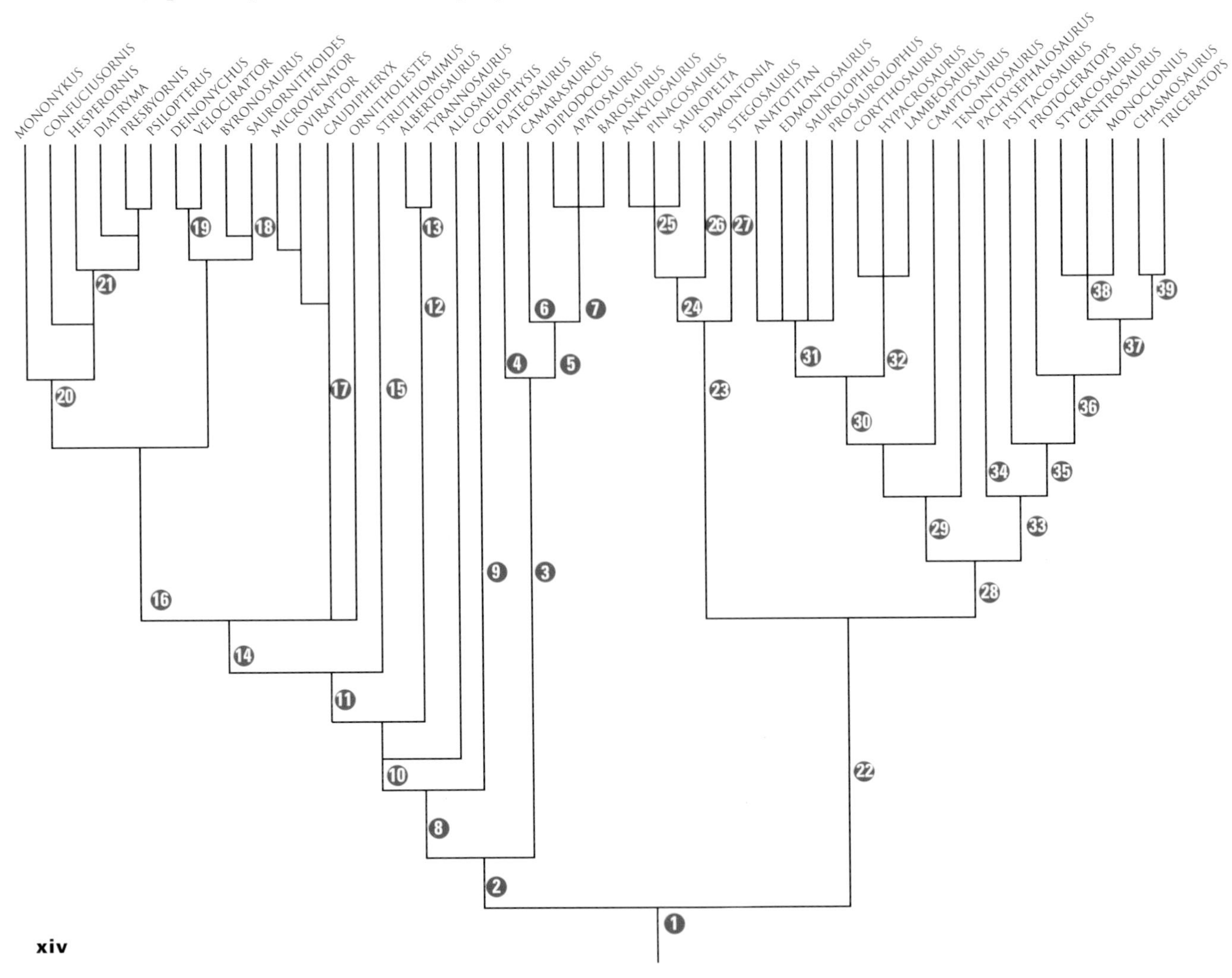

dinosaur groups, cladism has had a tremendous effect on our understanding of one group: birds. The now well-supported hypothesis that birds are the closest relatives of theropod dinosaurs has greatly affected interpretations of behavior and ecology of the extinct dinosaurs.

We present here a cladogram of many of the dinosaurs displayed in the American Museum of Natural History. Because the Museum's display of dinosaurs is the most complete and most extensive of its kind, this cladogram represents all the main groups of dinosaurs. The major features of the cladogram are based on the work of Paul Sereno of the University of Chicago, Jacques Gauthier of the California Academy of Sciences, Mark Norell and Peter Makovicky of the American Museum and Jim Clark of George Washington University.

The cladogram can be thought of as a graphic system of nested sets. Names keyed to numbers reflect groups within groups. (Not all of the groups for which names have been proposed are listed, only those that are well supported by character evidence.) Hence, *Triceratops* is ❶ a dinosaur, ㉒ an ornithischian, ㉘ a cerapod, ㉝ a marginocephalian, ㉟ a ceratopsian, ㊱ a neoceratopsid, ㊲ a ceratopsid, and ㊴ a chasmosaurine. A good way to think about this is to compare this to yourself. You are a mammal, a primate, an anthropoid (a higher primate), a hominoid (humans and apes), and a hominid (a human). Such a classification for dinosaurs is based on the kind of evidence below. This section is meant as a primer to place the names used in this book in a context that reflects the evolutionary history of dinosaurs.

In keeping with the critical point of view stressed here, we emphasize that this cladogram is not necessarily true. In systematics as in much of science, stability is ignorance, and we hope that the dinosaur cladogram of ten years from now will be better tested.

A PHYLOGENETIC CLASSIFICATION OF DINOSAURS

❶ DINOSAURIA

The derived characters that define the dinosaur clade (including birds) are related to locomotion. In dinosaurs the hind limb is swung under the body and moves in a fore and aft plane. These animals have an upright erect posture, similar to mammals. The pelvis is completely vertical, and the hip socket, or acetabulum, bears the weight of the articulation of the leg at the top of the hip socket rather than on its inside, as in most other tetrapods. The upper margin of the hip socket is enlarged into a shelf that supports the thigh bone, or femur. Thus a hip socket with a hole in the center and an enlarged bony rim on the upper margin is a derived character found only in dinosaurs.

❷ SAURISCHIA

The two derived characters of the saurischians are the grasping forefoot with strongly offset thumb and second finger the longest, and the long, flexible,

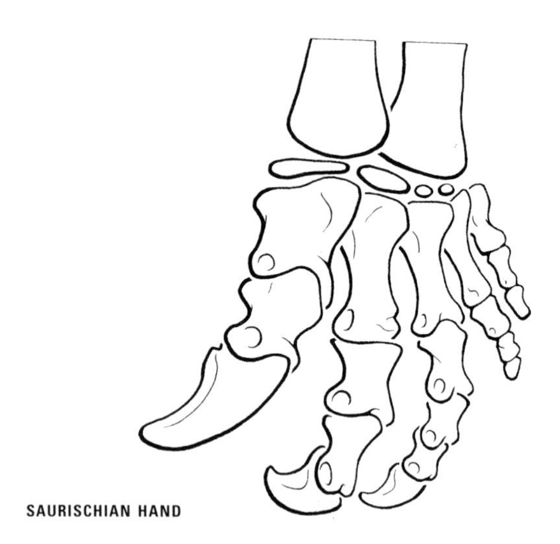

SAURISCHIAN HAND

S-shaped neck. The forefoot is distinctive in primitive saurischians, such as *Plateosaurus*, but highly modified in all later saurischians, such as sauropods, carnosaurs, and birds. Using this character to test the saurischian clade depends on accepting a series of transformations from a forefoot with five fingers and an offset thumb to highly modified forefeet such as the elephant-like feet of sauropods and the wings of birds. The S-shaped neck is obvious in most saurischians, although it is highly compressed in carnosaurs like *Tyrannosaurus*.

❸ Sauropodomorpha

Sauropodomorphs are united by only a few characters. All sauropodomorphs have small heads and long necks with at least ten neck bones, more than in other dinosaurs. Other tetrapods, however, have evolved small heads and long necks. Further work may show that this group did not have a unique common ancestor and is an artificial grouping.

❹ Prosauropoda

Because prosauropods are so primitive, determining unique characters to test the group is difficult. At present, the narrow, leaf-shaped teeth with serrated margins and the large thumb claws appear to be unique features of the group.

Plateosaurus

❺ Sauropoda

Sauropods form a well-defined group having a number of derived characters that show descent from a common ancestor. These characters include large spaces in the twelve or more neck bones, a reduced number of wrist and ankle bones, reduced forefoot fingers enclosed in a horseshoe-shaped pad, and large nostrils placed high on the skull.

❻ Camarasauridae

Members of the Camarasauridae are recognized by braincase features and by neck vertebrae with a divided U-shaped spine on top.

Camarasaurus

❼ Diplodocidae

This group is recognized by a long skull, nostrils that face directly upward, and a lightly built lower jaw with slender, peglike teeth.

Diplodocus
Apatosaurus
Barosaurus

❽ Theropoda

This well-tested clade represents tremendous diversity, from *Tyrannosaurus* to hummingbirds. The most obvious of the theropod ("beast foot") characters is the hind foot, which has only three functional toes (the middle toe is the largest).

⑨ Ceratosauria

This group includes the more primitive theropods but it may not represent a true clade. The principal investigators argue that neck bones with large air spaces, fused ankle bones, and other features of the backbone and pelvis are unique to this group.

Coelophysis

⑩ Tetanurae

This group consists of all but the more primitive theropods and is well defined by unique characters. The hand has lost the fourth and fifth digits, and the skull has unique characters, including the placement of all the teeth ahead of the eye socket. A long process (extension) attaches one of the ankle bones, the astragalus, tightly to the tibia.

⑪ Carnosauria

This group contains the largest theropods. They are united by a few characters, but further research may show that some members do not belong in the group. One of the characters that supposedly unites the group is the lacrimal foramen, a small hole in the bone directly in front of the eye.

⑫ Allosauridae

Relationships within the carnosaurs are not completely understood, yet various groups have been recognized. *Allosaurus* and a few other species are united into Allosauridae by a complex sinus in the upper jaw.

Allosaurus

⑬ Tyrannosauridae

Tyrannosaurus and its relatives have many unique characters. Among these are peculiar features of several bones in the skull, features of the neck and pelvis, and the famous character: only two fingers on the hand.

Tyrannosaurus

Albertosaurus

⑭ Coelurosauria

Coelurosaurs comprises the birds and their close relatives. The group is well tested by unique characters in the skull and neck and forelimb. The forelimbs are uniquely long, longer than half the length of the hind limbs, and have a thin, elongate hand with particularly long second and third fingers.

⑮ Ornithomimidae

The so-called ostrich dinosaurs are a well-defined group, with little diversity. The ornithomimids have a very lightly built skull with a long, shallow snout, and a very large eye socket and a derived fore and hind limb. Most have a toothless bill, but primitive forms have a few teeth.

Struthiomimus

⑯ Maniraptora

Cladistic analyses supports the old idea that birds descended from extinct dinosaurs, leading to the designation of Maniraptora as a clade. Birds and their closest relatives make up this group, which is defined primarily by features of the forelimb. The unique character that defines this group is a highly specialized wrist with a semilunate carpal, that is, a wrist bone in the shape of a half moon. The three fingers of the hand are also uniquely shaped. Although apparent in the oldest bird, *Archaeopteryx*, these features are modified in later birds.

Ornitholestes

Microvenator

⑰ Oviraptoridae

Oviraptorids are among the most unusual dinosaurs known. *Oviraptor* and all of its near relatives have high-domed skulls with large air sinuses similar to modern hornbills. The skulls terminate in unusual toothless beaks, which in life were probably covered by a horny bill.

Oviraptor

⑱ Tröodontidae

Trödontids are derived maniraptors that have a plethora of advanced characteristics. They have more teeth than any other maniraptor, and the teeth are recurved with very large serrations on their rear edges.

Saurornithoides

⑲ Dromaeosauridae

Dromaeosaurs were agile carnivores of the Mesozoic. This group is supported by several derived characters including a toe of the foot modified into a large retractable sickle claw, and powerful claws on the hand. The tails of dromaeosaurs have elongate, vertebral articulations that acted as stiffening rods, giving the tail a rigid construction.

Deinonychus

Velociraptor

⑳ Avialae

Avialians are the most derived theropods. Because they have living members, they can easily be characterized using advanced features. Avialians have a body covering of feathers and front appendages that are modified into wings. Other modifications of the shoulder and pelvic girdle also distinguish this highly diverse group.

Mononykus

Hesperornis

㉑ Aves

Aves comprises modern birds, which can be distinguished from more-primitive avialians by a number of derived features, including lack of teeth.

Diatryma

Presbyornis

Psilopterus

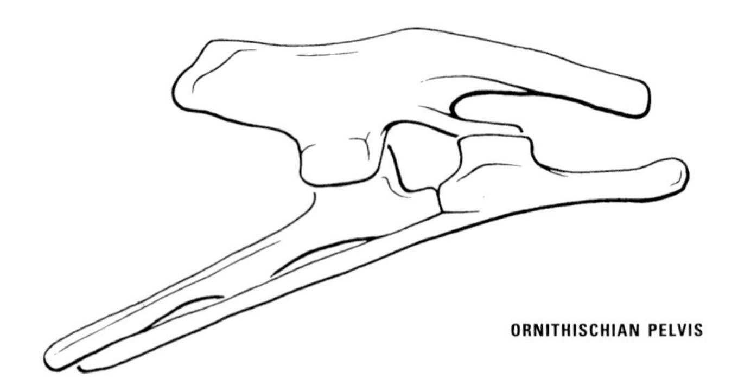

ORNITHISCHIAN PELVIS

㉒ ORNITHISCHIA

The ornithischians are the most diverse extinct dinosaurs and are characterized by horns, spines, plates, frills, and many other bizarre features. All are thought to have been plant eaters; most were quadrupedal ("four-footed"). The ornithischians are a well-tested clade, defined by the pelvis, which has a process (extension) from the pubic bones that points down and back along the ischium. The front teeth are small or absent and are replaced at the front by a horny beak. An extra lower jaw bone, the predentary, is unique to ornithischians and bears the beak. In advanced ornithischians the teeth are complex, but in primitive types they are leaf-shaped and triangular with a series of cusps, a tooth form unique to ornithischians.

(23) Thyreophora

The argument that armored dinosaurs, Ankylosauria and Stegosauria, have a unique common ancestor is based on only a few characters. Both possess armor plates that have keels arranged in rows on the upper surface of the body.

(24) Ankylosauria

The ankylosaurs are well-tested and distinctive, with many unique characters. They are extremely well armored with bony plates covering the flanks and neck. Among the unique characters are the fusion of many skull bones, the presence of a secondary palate that separates breathing and chewing in the mouth, and extensive modification of the pelvis.

(25) Ankylosauridae

The unique characters of ankylosaurids are their high-arched nasal regions with complex sinuses and unique tail club.

Ankylosaurus

(26) Nodosauridae

The nodosaurs have several unique derived features including an hourglass-shaped palate and a skull roof with a large platelike area between the eyes.

Sauropelta
Edmontonia

(27) Stegosauria

The stegosaurs, or plated dinosaurs, are a clade based on derived characters in the skull, tall arches in the backbone, and the very obvious upright armor on the midline.

Stegosaurus

(28) Cerapoda

Cerapods are united into one group by a series of unique characters: asymmetrical enamel on the cheek teeth which, causes uneven wear when chewing, as well as derived characters in the skull and pelvis.

(29) Ornithopoda

The ornithopods form one of the largest and most widespread Mesozoic dinosaur clade. This clade is well-tested with derived characters including the position of the lower jaw articulation with the skull. In other ornithischians the articulation lies on the same plane as the tooth row, but in ornithopods the articulation lies below the tooth row.

Tenontosaurus
Camptosaurus

(30) Hadrosauridae

The hadrosaurs, or duck-billed dinosaurs, were the most diverse and abundant large vertebrates during the last stages of the Mesozoic. The hadrosaurs are a well-defined group tested by unique derived features in the skull, the best of which is the complex tooth battery containing hundreds of teeth. The teeth are closely packed with three to five replacement teeth ever-growing in each tooth position to produce a compact, filelike surface for efficient crushing and grinding. The hadrosaurs are easily the best-known extinct dinosaurs.

(31) Hadrosaurinae

Hadrosaurines have relatively flat heads and solid spikes but very large nasal openings. A recent hypotheses of relationships suggests that some hadrosaurines may be more related to lambeosaurines and that Hadrosaurinae is an artificial group.

Prosaurolophus

Saurolophus
Edmontosaurus
Anatotitan

32 Lambeosaurinae

In contrast to hadrosaurines, lambeosaurines have greatly expanded hollow crests on the skull.

Corythosaurus
Hypacrosaurus
Lambeosaurus

33 Marginocephalia

This group is called Marginocephalia because the principal derived character involves the rear margin of the skull. Ceratopsians and pachycephalosaurs have an expanded frill or shelf at the back of the skull.

34 Pachycephalosauria

The pachycephalosaurs are a well-defined group with many shared derived characters. The most obvious is the greatly thickened bone in the skull roof, forming a prominent dome behind the eyes.

Pachycephalosaurus

35 Ceratopsia

Most of the horned dinosaurs, or ceratopsians, have prominent horns, but the primitive members of the group lack horns. The group is characterized by the presence of a unique bone at the front of the skull, the rostral bone, which supports the narrow, birdlike beak.

Psittacosaurus

36 Neoceratopsia

The principal derived character of neoceratopsians is the relatively large head with a prominent frill extending from the back of the skull.

Protoceratops

37 Ceratopsidae

This group represents the great diversity of advanced horned dinosaurs. Ceratopsids have large frills, well-developed horns, and a complex tooth battery. Other derived characters in the skull make this group a well-defined, well-tested clade.

38 Centrosaurinae

The derived characters of the centrosaurines are a short, high face and a relatively short frill.

Centrosaurus
Styracosaurus

39 Chasmosaurinae

The derived characters of the chasmosaurines are a long, low face with a long frill.

Triceratops
Chasmosaurus

1 QUESTIONS ABOUT DINOSAURS

Question #1

What are dinosaurs?

Popularly, "dinosaur" means any giant extinct "monster," especially ones that are ugly, stupid, and ill-suited to their environment, or anything else that is out of date. Scientifically, the group called dinosaurs must be defined in terms of evolution. This group must include all the animals descended from the first dinosaur, the common ancestor. Because the fossil record is incomplete and does not preserve all animals that have lived in the past, we will probably never find and identify the ancestor of all the dinosaurs. Nonetheless, we can recognize all the members of the group by identifying the unique evolutionary characteristics that they inherited from the first dinosaur.

FIGURE 5. The pelvis of the ornithischian *Thescelosaurus* (AMNH 5889) shows one of the derived characters of the Dinosauria (including birds). The hind leg or femur articulates in the circular area, the acetabulum, in the center of the pelvis. In the Dinosauria the acetabulum is open in its center as seen here, but in the primitive condition there is a bony wall.

Question #2

What are the closest relatives of dinosaurs?

Dinosaurs belong to a group called Archosauria. Archosaurs are divided into two groups, the Crurotarsi (crocodiles and their relatives) and the Ornithodira, a group composed predominantly of pterosaurs and dinosaurs. The living Archosauria are the twenty-one extant crocodiles and alligators, along with the more than ten thousand species of living theropod dinosaurs (birds).

The first person to examine the relationships among archosaurs using cladistics, Jacques Gauthier of Yale University, proposed that the closest relatives of dinosaurs were pterosaurs (the winged reptiles)

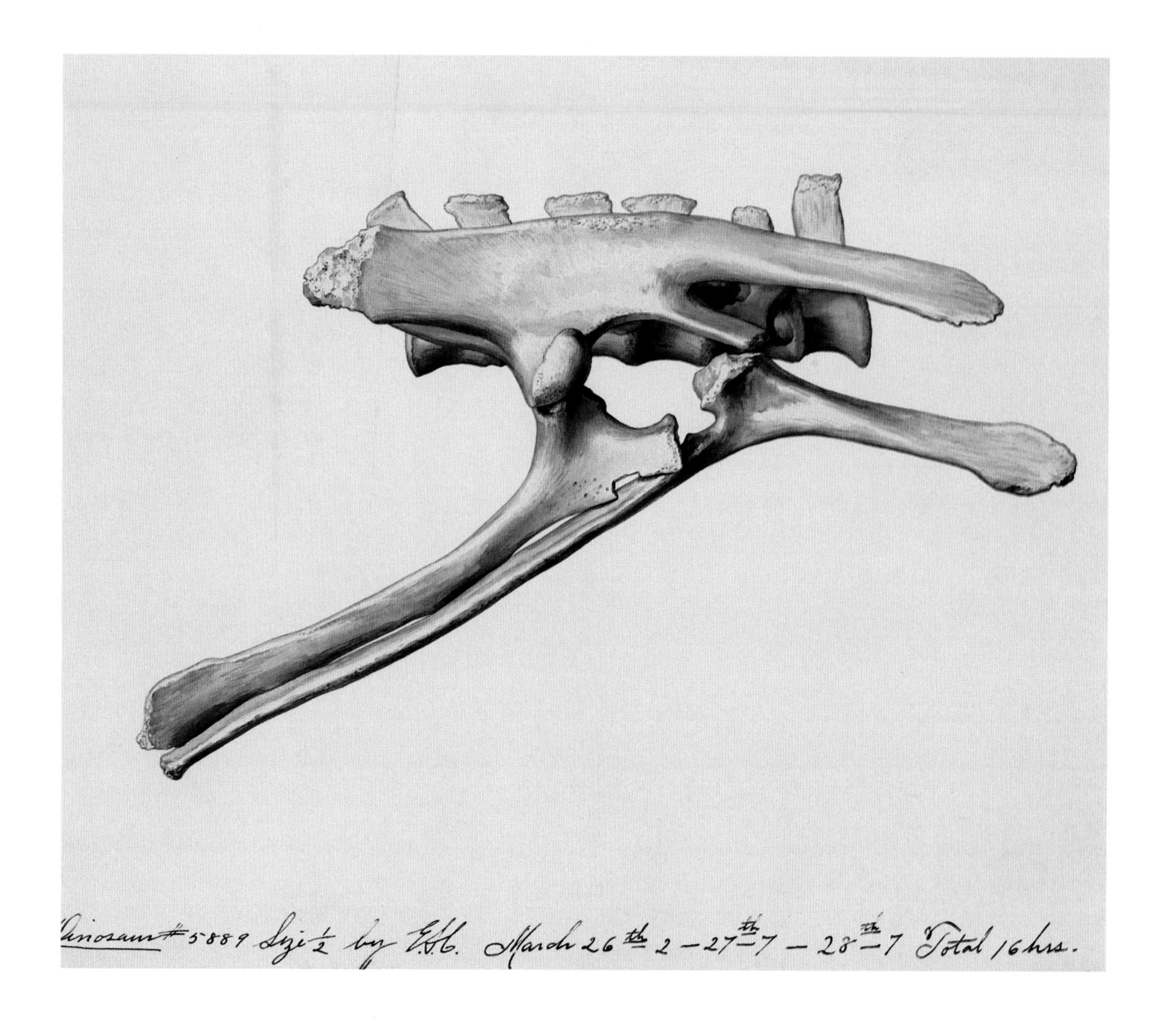

or *Lagosuchus* (a small bipedal archosaur). Recently this hypothesis has been refined to single out *Lagosuchus*. The argument is complicated by scientific nomenclatural difficulties with the type specimens. An appreciation for this problem requires some historical overview.

In 1971 the Harvard paleontologist Alfred Sherwood Romer named the genus *Lagosuchus* on the basis of a fragmentary specimen from the Middle Triassic Chañares Formation of La Rioja, Argentina. Romer recognized two

species, *Lagosuchus talampayensis* and *Lagosuchus lilloensis*. Specimens from the Los Chañares locality are fragmentary, and often more than one type of animal is found associated in a single nodule.

Reanalysis shows that Romer's diagnosis of the type species *Lagosuchus talampayensis* does not provide evidence differentiating it from other archosaurs. Therefore, even though the material attributed to the other species, *Lagosuchus lilloensis*, can be diagnosed, it can not be attributed to the invalid (nondiagnostic) genus *Lagosuchus* (see "How do dinosaurs get their names?," page 8).

Marasuchus has been proposed to replace the generic name "*Lagosuchus*" for the species *"Lagosuchus" lilloensis*. The closest relative of dinosaurs is now considered to be *Marasuchus lilloensis*, which used to be called *Lagosuchus lilloensis*.

FIGURE 6. Animals similar to this reptilelike mammal are often called dinosaurs, but this animal is more closely related to humans than it is to dinosaurs. This early mount shows *Edaphosaurus* with the head of *Dimetrodon*, a related form. New evidence has shown that the actual head of *Edaphosaurus* was much smaller and lacked the large sharp teeth.

Question #3

Why are dinosaur fossils so interesting?

FIGURE 7. Dinosaurs are and were magnificent creatures. Many extinct dinosaurs, such as this *Chasmosaurus kaiseni* (AMNH 5401), represent body types that are very different from present-day animals.

From toddler to octogenarian, almost everyone is interested in dinosaurs. It is no wonder that the dinosaur halls are the most popular attraction of the Museum, but why these animal fossils have nearly universal appeal is hard to fathom. Scientists, psychologists, educators, and talk-show hosts have speculated on this issue. Here we present our own, somewhat quixotic view.

Humans have fertile imaginations. Even (and especially) as children, we respond to environmental cues, using our imaginations to define our place in nature. For many children dinosaurs are a natural starting point for imaginary journeys. Because many of the well-known dinosaurs are huge, some fearsome, and all extinct, they epitomize the concept of time and impart a sense of temporal place, somewhere in a distant past very different from today. They allow us to think about something that was real but beyond our household doors, and unlike anything in today's world. Naturally we want to know everything we can about these intriguing animals.

Dinosaurs are interesting because they are natural subjects for active imaginations, yet precisely this attribute has hindered the application of rigorous scientific methodology to the group. Many dinosaur paleontologists have been enticed to apply their vivid imaginations to problems, rather than to restrict themselves to using the objective tools of empiricism and logic to test their ideas scientifically. Nonetheless, dinosaur fossils represent a rich data set that can be used to test ideas about a broad range of evolutionary problems, many of which we discuss in this book. These problems, more than the animals themselves, are what scientifically interest paleontologists.

Question #4

Who discovered the first dinosaur bones?

In the past, people had far more intimate knowledge of the construction of vertebrate bodies than we do now. This knowledge probably began to develop when the first humans or human ancestors began making tools and butchering animal carcasses for food. Such anatomical familiarity would have continued or even increased as humans began to raise animals for food. It is reasonable to assume that peoples with an intimate knowledge of the construction of vertebrate bodies would recognize the remains of fossil animals. Collecting practices thus probably have ancient roots. In contemporary Chinese markets bones of dinosaurs and other fossil vertebrates are readily available for use in traditional medicine. Although not called dinosaur bones, they are recognized as the fossil remains of long-dead animals, often identified as remains of dragons. Chinese pharmacopoeias at least as old as the Han Dynasty refer to these ingredients, and it is not unreasonable to posit that the Asian practice of collecting fossil bones for medicinal purposes significantly predates this written record.

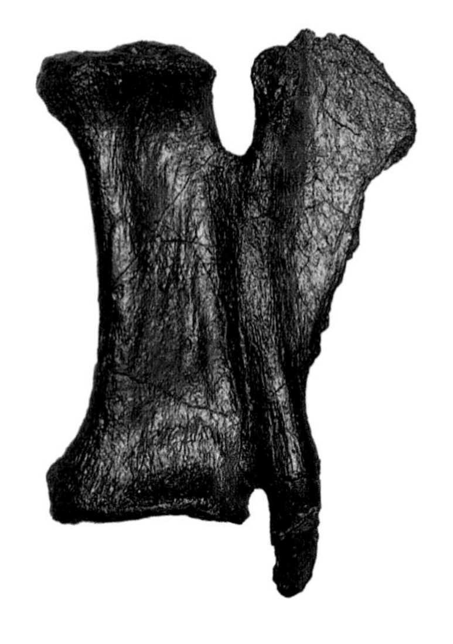

FIGURE 8. This specimen (AMNH 223) is part of the vertebral column of *Diplodocus*, a large sauropod. Collected by Barnum Brown in 1897 at Como Bluff, Wyoming, it was the Museum's first dinosaur specimen.

In northern Europe Robert Plot made the first unmistakable reference to dinosaur bones in 1677. Plot described the lower end of the thigh bone that formed the knee of *Megalosaurus,* a bipedal carnivorous dinosaur common in the Jurassic rocks of southern England. Plot did not recognize his find as the bones of a long-dead giant reptile. Instead (so the story goes), he believed he had found the fossilized testicles of a giant antediluvian man.

The first dinosaur fossils recognized as fossilized elements of giant extinct reptiles were found in 1824 by the English geologist and minister William Buckland, of Oxford University. Buckland collected the jaws of a large carnivorous dinosaur, and recognizing its reptilian affinities, named it *Megalosaurus.* Two years earlier, Mary Ann Woodhouse Mantell had found the teeth of a large plant-eating dinosaur during a road-building project near her house in Sussex, England. She gave these teeth to her husband, Gideon Mantell, a local physician and amateur paleontologist. At the Hunterian Museum in London the teeth were examined by many of the great comparative anatomists of the day, including Georges Cuvier, who concluded that they were the teeth of a prehistoric herbivorous reptile. Noting the similarity (except in size) between the teeth found by Mary Ann Mantell and those of the extant, herbivorous lizard *Iguana*, Gideon Mantell named the second formally recognized dinosaur *Iguanodon* in 1824.

By the early 1800s, bones of dinosaurs and other fossil vertebrates had begun to appear in museum collections in North America, but the realization that they were the remains of large extinct reptiles came only after European reports. The first documented "find" in North America appears to have been reported by a Philadelphia anatomist Caspar Wistar, who described a giant leg bone of what probably was a hadrosaur. This specimen was not recognized as a fossil reptile and has been lost in the intervening years.

Dinosaur science in North America formally began in the 1850s, when Philadelphia zoologist Joseph Leidy and geologist-explorer Ferdinand Hayden described *Troödon* and *Paleoscincus* from the American West and *Hadrosaurus* from New Jersey.

Question #5

Who coined the term "dinosaur"?

The term "dinosaur" was coined by the British anatomist Sir Richard Owen in 1841. It is derived from the Greek *deinos* ("terrible") and *sauros* ("lizard"). In 1887 British geologist Harry Govier Seeley divided dinosaurs into two groups, the saurischians and ornithischians (*see* "How many different kinds of dinosaurs are there?," page 15).

Question #6

How do dinosaurs get their names?

Naming an organism, or determining an organism's correct name, is regulated by a set of rules called the International Code of Zoological Nomenclature (ICZN). Like all other animals, dinosaurs are named according to these rules. Although the ICZN rules may seem restrictive or unnecessary, they are important because they establish a standard, so paleontologists (and other biologists) can devote time to more interesting topics than naming organisms.

The ICZN stipulates that all names must be constructed as binomials; that is, each name be composed of two parts. The first part, the genus, is always capitalized; the second, the species, is never capitalized, even when derived from a proper noun. Both names are always italicized, and sometimes the genus name is abbreviated (for example, *T. rex*). Hence the correct name for a given specimen of *Apatosaurus* is *Apatosaurus excelsius, Apatosaurus louisi,* or *Apatosaurus ajax*, depending on the species. The genus may be used alone to refer generally to all species of the group.

The first person to describe a new type of fossil assigns the animal its name. The root of the name can come from any source. Often these names are derived from Greek or Latin words. Many names reflect an outstanding characteristic of the organism, the place where the fossil was collected, or the person who collected it.

The original name is tied directly to a particular fossil, the type specimen. The specimen must exhibit features that distinguish it from any other species; any new specimens thought to belong to this species must be found to have no significant differences with the type specimen.

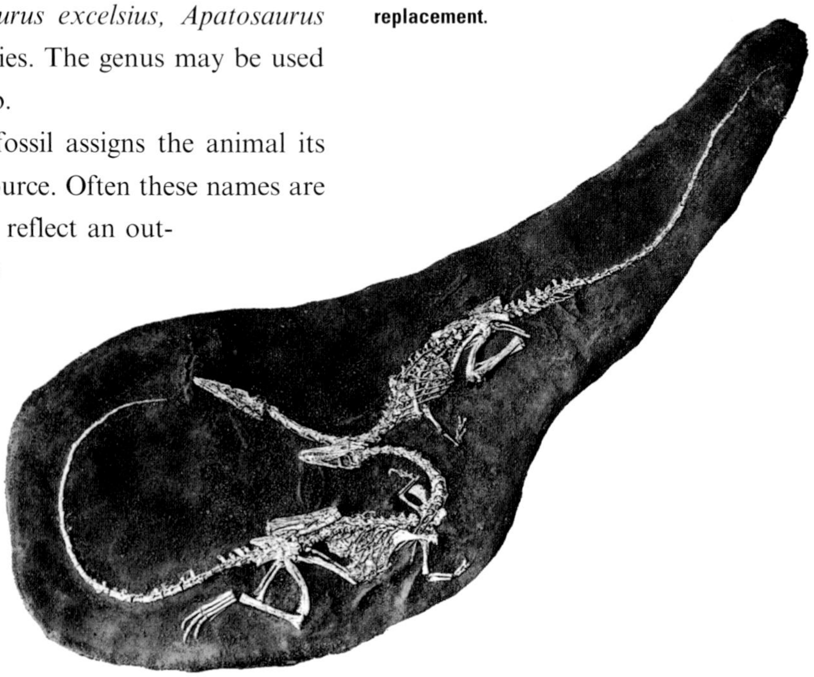

FIGURE 9. The dinosaur *Coelophysis bauri* (AMNH 7223 and 7224) is currently at the apex of a nomenclatural debate. The type specimen was collected in 1881 by David Baldwin. Reanalysis of the type specimen has indicated that it may not be diagnosable—that is, it may not preserve characteristics that separate it from all other known animals. If so, the name *Coelophysis bauri* is invalid. The name *Rioarribasaurus colberti* has been suggested as a replacement.

FIGURE 10. The American Museum of Natural History's *Apatosaurus*, the first large dinosaur mounted for exhibition, was originally called *Brontosaurus*. Later *Brontosaurus* was interpreted to be the same species as the earlier-named *Apatosaurus*. Since *Apatosaurus* was the first name used, it is the official name.

Two nomenclatural rules—priority and synonymy—have important ramifications in dinosaur science. In 1993 a small Mongolian theropod was named *Mononychus olecranus*. The name *Mononychus* ("one finger") refers to the single large claw on its hand. The species name, *olecranus*, refers to the olecranon, or large bony prominence at the elbow. After the name was assigned, it was discovered that the name *Mononychus* had already been used for a small beetle described in the early 1800s. Because *Mononychus* was preoccupied (the principle of priority), the name had to be modified. The investigators chose to replace the "ch" with a "k," thus *Mononykus*. This modification is allowed because even though the names may be pronounced the same, they are not spelled exactly alike.

A more famous example, which illustrates the principles of synonymy and of priority, is the case of *Brontosaurus*. In 1877 the Yale paleontologist

O. C. Marsh described a fine specimen of a sauropod from Como Bluff, Wyoming, that included most of the back vertebrae, several limbs, and a good bit of the tail. His paper, entitled "Notice of a New Gigantic Dinosaur," appeared in the *American Journal of Science* without illustration. Marsh named this sauropod *Apatosaurus ajax*. Two years later Marsh published another paper, "Notice of New Jurassic Reptiles," describing a very good sauropod skeleton. Marsh had several of the bones of this second specimen, which he named *Brontosaurus excelsius*, illustrated. Probably because of these illustrations and the catchy name, *Brontosaurus* became ingrained in both popular culture (consider the brontoburgers in the cartoon series "The Flintstones") and in the scientific literature.

After Marsh's death in 1899, Samuel Williston, a coworker of Marsh, recognized that the bones of *Apatosaurus ajax* and *Brontosaurus excelsius* actually belonged to the same type of dinosaur. *Apatosaurus* and *Brontosaurus* were formally designated synonyms in 1903. Obeying the rule of priority, *Apatosaurus* is the valid name because it was published first, even though *Brontosaurus* is the commonly used, but incorrect, name.

Problems involving the rules of priority and synonymy are common. Many of these difficulties stem from the designation of the original type specimens. At this writing, a nomenclatural debate regarding the dinosaur *Coelophysis bauri* (*see* page 107) is raging. *Coelophysis* was originally collected in 1881 and described in 1889. In the 1940s the Museum collected many new specimens of this animal at Ghost Ranch, New Mexico. Recent study of the original specimens has indicated to some paleontologists that the specimens are not diagnosable, meaning that the specimens are so fragmentary that characteristics establishing that they are from a unique species are not present. If new specimens are collected, they cannot be referred to the original type material, because the type material is so poor that it preserves no unique characteristics. Undiagnosable type material renders a name a *nomen dubium* (that is, an invalid name). One suggestion in the debate over *Coelophysis* has been to designate one of the excellent specimens from Ghost Ranch as the type specimen, with the new name of *Rioarribasaurus colberti*, referring to the county where the type was collected (Rio Arriba) and American Museum of Natural History curator emeritus Edwin Colbert. Although the debate continues, eventually scientists will accept either *Coelophysis* or *Rioarribasaurus* as the official name.

Question #7

Why are birds a type of dinosaur?

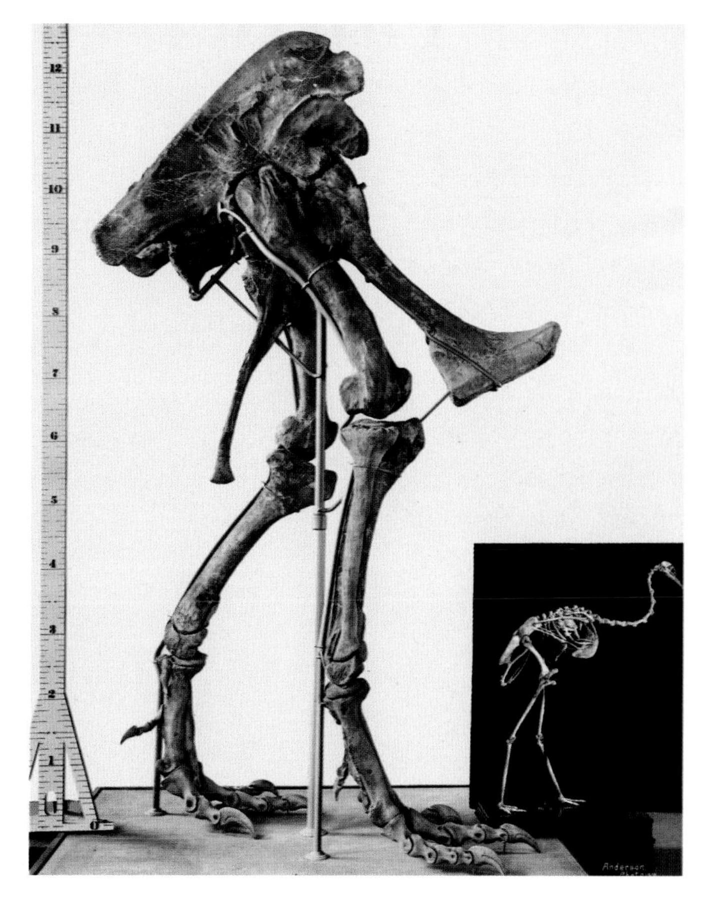

FIGURE 11. Many reasons that birds seem so different from other dinosaurs relate to scale. Many skeletal features are very similar. For instance, compare the leg structure of *Tyrannosaurus rex* with this wading bird. Both animals have three primary digits on each foot.

The idea that the origin of birds is intimately involved with the evolution of dinosaurs goes back to the Victorian era and two of the great scientists of nineteenth century, T. H. Huxley and Richard Owen. Studies by these men on early dinosaur finds and on the archetypal primitive bird, *Archaeopteryx*, discovered in 1869, documented anatomical evidence that firmly supported a dinosaurian origin for birds. According to one story, one Christmas Day Huxley was carving a turkey for his annual feast. As he dissected the drumstick he was struck by an unmistakable similarity between his Christmas dinner and the fossils of the theropod *Megalosaurus* back in his office. From that day on Huxley proclaimed (to use his own stiff Victorian words), *"...surely there is nothing very wild or illegitimate in the hypothesis that the phylum of the class Aves has its roots in the dinosaurian reptiles."* Whether or not this story is true, the evolutionary link between birds and extinct dinosaurs has been anatomically recognized for nearly 125 years.

This theory has not always been popular. Since the studies by Owen and Huxley, many in the paleontologic elite have sought the origin of birds elsewhere, usually in the poorly defined group of primitive reptiles called thecodonts. The collection of new fossil material and the development of cladistic analysis, however, have transformed the way we look at birds.

Contemporary studies have indicated not only that birds are related to dinosaurs but that they are a type of dinosaur, specifically a type of advanced

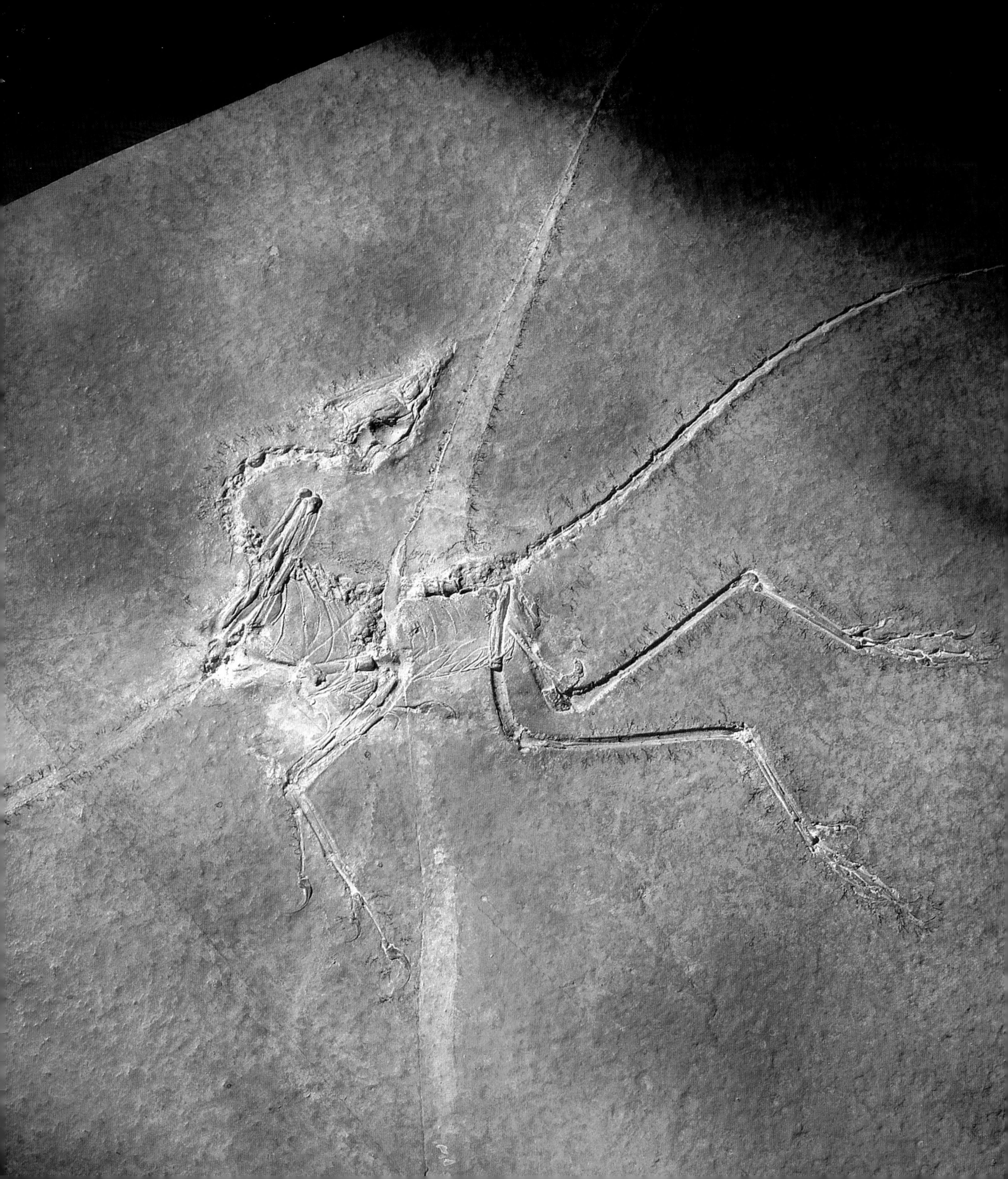

FIGURE 12. ***Archaeopteryx lithographica*** **from the Jurassic of Germany is the oldest and most primitive known bird. Its long bony tail and semilunate carpal in the wrist are features derived from more primitive dinosaurs; its feathers are derived features of birds. *Archaeopteryx* is both a bird and a dinosaur in the same way that humans are both primates and mammals.**

theropod called a maniraptor. Maniraptors are a group of small agile theropods that includes birds, as well as dinosaurs like *Ornitholestes, Deinonychus, Velociraptor*, and *Saurornithoides*. Thus, in the same way that humans are primates and primates are mammals, birds are dinosaurs because they share unique evolutionary characteristics found only in other dinosaurs and therefore, evolved from the common ancestor of dinosaurs. From here on the word "dinosaur" will refer to birds as well as the extinct giants of the past. When we refer to what comes to mind when most people think of dinosaurs, we will be careful to say "nonavian dinosaurs" (all the dinosaurs except birds).

The contemporary perspective on the origin of birds began to develop in the mid-1960s. About this time John Ostrom, a paleontologist at Yale University, made the first in a series of important discoveries improving our understanding of bird origins. Ostrom's first discovery involved the spectacular specimens of *Deinonychus* in Early Cretaceous rocks of southern Montana. Although legendary Museum dinosaur hunter Barnum Brown had collected specimens of this animal more than thirty years earlier, he had never completed and published a scientific description (*see* "*Deinonychus*," page 129). Ostrom's description of *Deinonychus* in 1969 was to be crucial in the recognition of the closest evolutionary relatives of birds because this animal shares many unique features with primitive birds like *Archaeopteryx*.

Ostrom's second discovery came during a visit to European museums in which he examined specimens of *Archaeopteryx* and documented the anatomical evidence linking *Archaeopteryx* and *Deinonychus*. For instance, the forelimbs and hind limbs of *Archaeopteryx* are smaller versions of those in *Deinonychus*. This realization enabled Ostrom to discover the Haarlem specimen of *Archaeopteryx* in the Netherlands Teyler Museum. Unlike most other known specimens, this specimen preserved only faint feather impressions with the skeleton. But there was no mistaking it. The shape and proportion of bones in the forearm were identical to those of other *Archaeopteryx* specimens and very similar to those of *Deinonychus*. This specimen, previously labeled a pterosaur, was another *Archaeopteryx* specimen.

The next great advance in our understanding of bird origins was conceptual, the advent of cladistic analysis and its empirical basis for reconstructing phylogeny (*see* "Introduction," page 10, and "The Evolutionary Relationships of Dinosaurs," page xiv). Using cladistics, Jacques Gauthier, of Yale University, evaluated the evidence for relationships among theropod dinosaurs. On the basis of a wide variety of characteristics, Gauthier hypothesized that birds were the sister group (the closest relatives) of a group of maniraptoran theropods that included troödontids (for example, *Troödon* and *Saurornithoides*) and dromaeosaurs (*Deinonychus* and *Velociraptor*).

Since Gauthier's work in the early 1980s, important new fossil remains have come to light, including the primitive birds *Sinornis, Patagopteryx, Iberomesornis*, and *Confuciusornis*, and interesting nonavian maniraptors (such as the Chinese troödontid *Sinornithoides*, and more complete and informative specimens of *Velociraptor* and *Oviraptor*). The new animals that have been collected, combined with new evidence from the reexamination of old specimens, support the hypothesis that birds are a type of dinosaur, specifically that they are maniraptors. This hypothesis is still being refined, and work on this problem is a primary concern of dinosaur paleontologists.

What is the evidence that birds are closely related to these small agile theropods? All maniraptors, including birds, share unique evolutionary characteristics inherited from their common ancestor, including specializations of the shoulder, forearm, and wrist. For example, a crescent-shaped bone in the wrist (called a semilunate carpal) is found only in maniraptors. In maniraptors like *Velociraptor*, this bone articulates with (fits up against) the bones of the hand, with its crescent-shaped surface forming part of the wrist joint. In birds this bone has become fused to the wrist but it retains its characteristic shape, and its common presence in birds and animals like *Velociraptor* is evidence for a close evolutionary relationship. Several more obscure anatomical features found uniquely in maniraptors support this relationship, including the construction of the pelvis, the development of large air spaces in the skull, and the detailed structure of feet and vertebrae. At a more general level on the cladogram dinosaurs as different as pigeons and *Tyrannosaurus rex* share some typical theropod features inherited from the common ancestor of all theropods. The S-shaped curve of the neck, as well as the foot composed of three primary toes with a small fourth toe held high off the ground on the back of the foot, are indicative of the pigeon's dinosaurian heritage. The great majority of available anatomical evidence supports the hypothesis that birds evolved from the common ancestor of all dinosaurs.

Question #8

How many different kinds of dinosaurs are there?

The fossil record provides our only direct insight into the diversity of extinct dinosaurs. Because the fossil record is incomplete, we can view the record only as a minimum estimate of the number of kinds of dinosaurs that existed in the past. The completeness of the fossil record depends on the abundance of animals in life and the fossilization potential of the environment that the animals lived in—both subjects that we will examine later.

We do not know how bad the dinosaur fossil record is, but we suspect it may be very bad! Consider, for example, the Late Cretaceous (about 70 million years ago) and the Middle Jurassic (about 175 million years ago). If we tally the known dinosaur types from each period, we see that many more kinds of dinosaurs have been found in Late Cretaceous rocks than in Middle Jurassic rocks. Were there really more types of dinosaurs in the Late Cretaceous than in the Middle Jurassic? We can't tell, because our comparisons are not equivalent. For example, many more Late Cretaceous localities have been discovered and are producing fossils of more kinds of dinosaurs than are Middle Jurassic localities. Thus, it is not surprising that we know about more types of dinosaurs from the Late Cretaceous, but the difference may be simply that not many Middle Jurassic localities have been discovered.

Also influencing estimates of extinct dinosaur diversity is the geographic distribution of fossil localities. Radically different sorts of animals are restricted to specific continents. Consider antelopes of today's Africa, North America's bison and elk, and Australia's kangaroos. Undoubtedly, different dinosaurs lived on different continents, a fact borne out by the known fossil record. Therefore, time periods that have fossil localities distributed all over the world should have more kinds of dinosaurs. Late Cretaceous faunas are distributed across the planet, but Middle Jurassic faunas are known from only a few places; from this fact alone we would expect to know less about the diversity of Middle Jurassic dinosaurs than of Late Cretaceous dinosaurs.

FIGURE 13. American Museum of Natural History and Mongolian field crews collecting new specimens of dinosaurs in Mongolia in 1991. Because of the rarity of dinosaur fossils, new collections often include new species.

These kinds of inequalities in the number and distribution of dinosaur localities are called sampling biases. If we fail to recognize these biases, we can propose some thought-provoking claims. For example, today at least ten thousand bird species populate our planet; some recent estimates place the number nearer to twenty thousand. A literal reading of the fossil record, however, indicates that during the heyday of dinosaur diversity (that is, the limited time in the Late Cretaceous from which most species are known) there are only about forty species of dinosaurs known worldwide. Contemporary bird diversity is 250 times greater than that found in the best-sampled interval of the Mesozoic. Uncritical examination of the fossil record leads to the conclusion that dinosaurs are hundreds of times more diverse today than they ever have been—a proposition that is hard to believe.

Qustion #9

What is the earliest-known dinosaur?

FIGURE 14. Because the fossil record is so incomplete, we cannot identify the first dinosaur. Most of the earliest dinosaurs that have been found were small, bipedal carnivores that looked similar to this *Coelophysis bauri*.

The best-documented candidates are from the Late Triassic Ischigualasto Formation of Argentina. Fossils found in these beds include the theropods *Herrerasaurus* and *Eoraptor*, as well as the ornithischian *Pisanosaurus*. The Ischigualasto beds contain a volcanic ash that has recently yielded a radiometric date of 228 million years ago (*see* "How do we estimate the age of dinosaur fossils?," page 81).

Other possible candidates include dinosaur remains from the Late Triassic of the Petrified Forest in Arizona, the Wolfville Formation in Nova Scotia, and some rock formations in Brazil. Because volcanic rocks are absent from these formations, however, no radiometric dates have been acquired. Instead, paleontologists have had to rely on biostratigraphic correlation (using pollen fossils or vertebrate remains). The evidence collected so far suggests that dinosaur remains from these localities are slightly younger than the *Eoraptor, Pisanosaurus*, and *Herrerasaurus* remains from the Ischigualasto Formation.

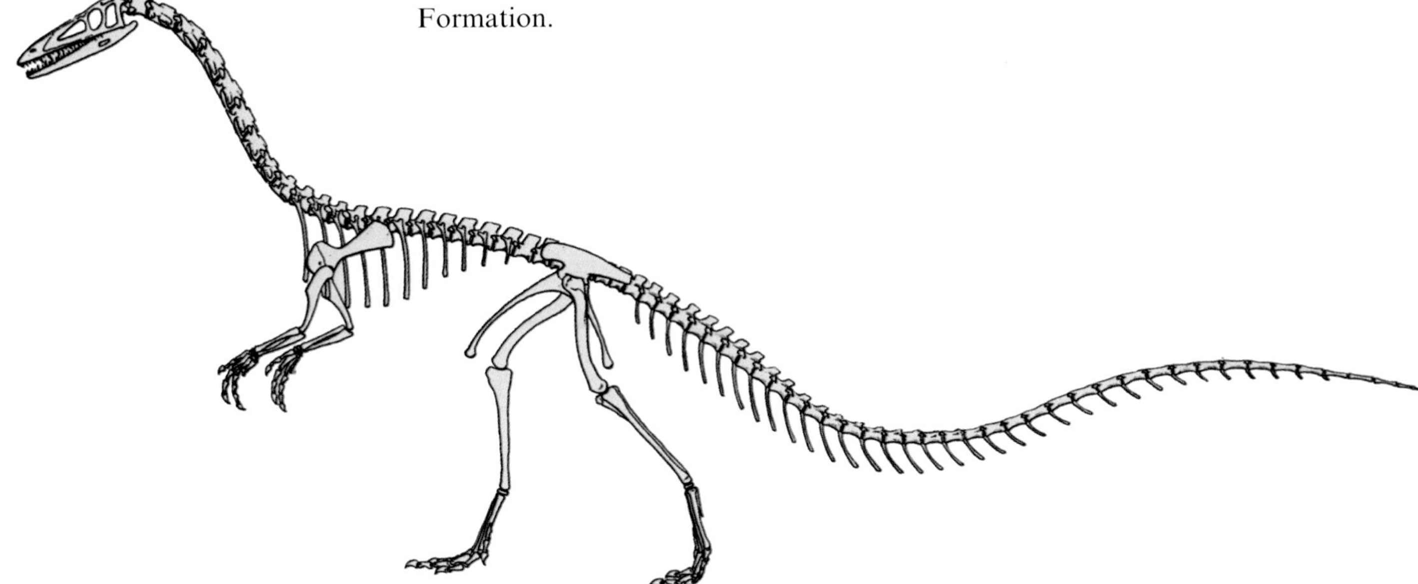

Question #10

Where did dinosaurs live?

Fossil bones and tracks of dinosaurs have been collected almost everywhere that terrestrial Late Triassic, Jurassic, or Cretaceous rocks are exposed. Dinosaur bones have been excavated on every continent and have been recovered from large continental islands such as Japan and Madagascar. Because they were (and are) such a diverse group, they probably lived in nearly every terrestrial environment. Fossil remains have been found in ancient habitats that appear, on the basis of geologic interpretation, to have represented deserts, savannas, forests, beaches, and swamps.

Dinosaur fossils have even been found in Antarctica and in North America above the Arctic Circle. Did the dinosaurs actually live in polar conditions? Remember that during most of the Mesozoic the world was a very different place. The global climate was generally much warmer than it is today, and giant polar ice caps probably did not exist. The positions of the continents today are very different from the positions of the Mesozoic continents (*see* "Earth History Timeline," endpapers). Studies of paleogeography tell us that some dinosaurs (whose remains have been found in Alaska, Antarctica, and, surprisingly, Australia) lived very near the poles. Even if the climate was much warmer, during certain periods of the year dinosaurs would have experienced extensive periods of darkness, which would have drastically influenced the food supply of the herbivores. How they may have compensated is a mystery, and a variety of mechanisms, such as hibernation or long migration, have been proposed to explain how they might have survived. Data to test these hypothetical mechanisms are difficult to gather, and for now these explanations remain speculation.

FIGURE 15 (OPPOSITE). Because dinosaurs have existed for so long, their history encompasses the entire history of most present-day plant and animal groups, as well as many that are extinct. This palm leaf, collected by R. T. Bird from the Picture Room of the Red Mountain Mine (*see* "The Coal Mine Tracks", page 179), is from the Late Cretaceous. It emphasizes that at least during the last stages of the Mesozoic era, many of the same sorts of animals and plants that exist today cohabited with dinosaurs.

Question #11

What was the world like during the time of the dinosaurs?

The nonavian dinosaurs flourished for over 170 million years. Given this extensive geographic and temporal distribution, dinosaurs probably encountered almost every climatic regime that the planet has to offer. Few generalizations about the climate during the Mesozoic can be made, except to say that it was probably milder than today. Paleoclimatologists call this a more equable climate. It is important to realize that the early stages of modern human history occurred during an ice age. Now we are in an interglacial stage; that is, we are between major glaciations (the previous one ended about 10,000 years ago).

The biggest difference between the Mesozoic and the present is not so much the kinds of animals present (most of the major groups living today had already originated in the Mesozoic), but the kinds of plants. Even in the final phases of the Cretaceous at the end of the Mesozoic, forests and prairies would have appeared very different from those of today. Several kinds of flowering plants, including many that are ubiquitous members of contemporary ecologic communities, such as grasses and most other advanced flowering plants, are not found as Mesozoic fossils. Hence, even if present, these plants were not dominant members of ecologic communities like they are today. Instead, Mesozoic plant communities were dominated by coniferous trees, gingkoes, and cycads.

Question #12

Did dinosaurs really rule the world?

Today there are nearly as many species of birds as of mammals, but we would never say that birds rule the planet. If we were to think of a group of organisms as "ruling the planet" (not a phrase commonly used in scientific discourse), we would probably measure the success of a group by its morphologic diversity (number of body plans), or by the extent of its impact on the environment. From our anthropocentric point of view, humans and our own mammalian relatives are the dominant life form. Although a good case could be made for insects or even flowering plants, two groups with mind-boggling numbers of species and varied morphologies, most humans are more impressed by large vertebrate animals, such as dinosaurs.

For almost 170 million years, from the Late Triassic to the end of the Cretaceous, there existed dinosaurs of almost every body form imaginable: small carnivores, such as *Compsognathus* and *Velociraptor*, ecologically equivalent to today's foxes and coyotes; medium-sized carnivores, such as *Deinonychos* and the troödontids, analogous to lions and tigers; and monstrous carnivores with no living analogs, such as *Tyrannosaurus* and *Allosaurus*. Included among the ornithischians and the elephantine sauropods are terrestrial herbivores of diverse body form. By the end of the Jurassic, dinosaurs had even taken to the skies (*see* "Did dinosaurs fly?," page 59). The only habitats that dinosaurs did not dominate during the Mesozoic were aquatic. Yet, there were marine representatives, such as the primitive toothed bird *Hesperornis*. Like penguins, these birds were flightless, specialized for diving, and probably had to return to land to reproduce. In light of this broad morphologic diversity, dinosaurs did "rule the planet" as the dominant life form on Earth during most of the Mesozoic.

Question #13

What other animals lived during the time of the dinosaurs?

FIGURE 16. The Cretaceous mammal *Zalambdalestes*, whose skull was found in Mongolia by an American Museum of Natural History field crew, cohabitated with dinosaurs such as *Protoceratops* and *Velociraptor*. Mammals in the Cretaceous were similar to, but not necessarily closely related to, shrews and rodents.

The fossil record documents that dinosaurs shared the Mesozoic Earth with a wide variety of other animals. Surprisingly, these include representatives from most of the major groups of animals alive today, many of which appeared before or at about the same time as the appearance of the first dinosaur. The earliest mammals appeared in the Late Triassic, along with the first turtles, frogs, lizards, and crocodiles. Consequently, all these groups lived with the dinosaurs until the disappearance of the nonavian dinosaurs, at the end of the Cretaceous.

By the end of the Cretaceous, modern birds, snakes, primitive marsupials, and placental mammals, as well as modern sorts of crocodiles, turtles, and lizards were all present. The fossil record is not good enough to tell us if the diversity of these groups matched what we see in similar ecosystems today, but we have no reason to believe otherwise (*see* "How many different kinds of dinosaurs are there?," page 15). If we could travel in time back to a late Mesozoic forest, in what is now western North America, we would find about as many different kinds of animals as we see today, most of them would appear ordinary, and many (with the exception of the nonavian dinosaurs) would be familiar.

Question #14

How large were the biggest dinosaurs?

FIGURE 17. Illustrating the size of the largest dinosaurs, the sauropods, a Texas ranch child, Tommy Pendly, sits in the water-filled track of one of the Paluxy River sauropod tracks during R. T. Bird's 1939 excavation of the specimen for the Museum.

Some dinosaurs reached extraordinary size. These land leviathans were the largest terrestrial animals ever to have lived. Determining which dinosaur was the biggest and how huge it was is difficult because many of the largest dinosaurs are known only from fragmentary remains. For others only a few skeletons have been completely excavated. Nonetheless, the title of "largest land animal ever" undoubtedly belongs to a member of the Sauropoda (the group containing *Apatosaurus*).

To discuss size, we need to establish what "large" means. Does it mean the longest dinosaur, the tallest, or the heaviest? Estimating the length and height of these animals is easy, if a fossil skeleton is reasonably complete. Weight is another matter. Consider two people: an average middle-aged, out-of-shape paleontologist, and a professional athlete. If we had only their skeletons we would have difficulty estimating their weights accurately because the amount of flesh on any given skeleton may vary by a factor of two. The same is true of fossil animals. From skeletons, we know that some weighed more than others (*Apatosaurus* and *Diplodocus* were nearly the same length, but *Apatosaurus* probably weighed more, because its bones were much more robust), but because we don't have a good idea of how much soft tissue hung on the fossilized skeletons, we have no real empirical data with which to determine the exact weight of the animal.

We estimate the weights of fossil animals by one of two methods, both woefully inadequate. The first is an experimental procedure: We construct a small scale model of a particular dinosaur in its live state, using the fossil skeleton as a guide. Then we measure its volume. Because the densities of almost all vertebrate animals are fairly uniform (nearly equal to water, at 1 gram per cubic centimeter), we can calculate the weight. The

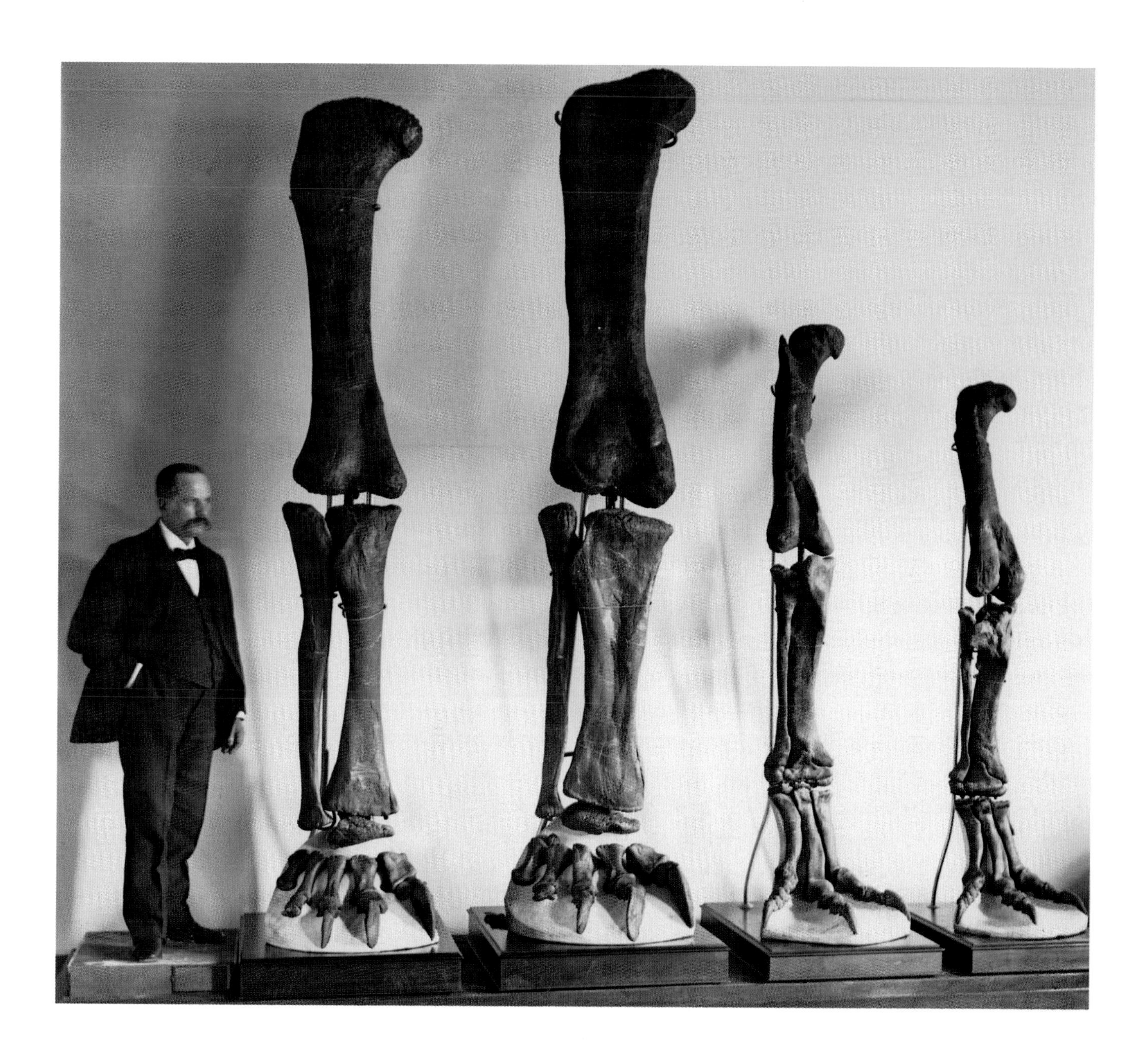

FIGURE 18. One of the first fruits of the Museum's dinosaur-collecting expeditions was a series of hind limbs of Jurassic dinosaurs. These included some of the largest dinosaurs known at the time. Techniques for mounting these bones were in their infancy, but people like Adam Hermann, shown above with (left to right) *Diplodocus, Apatosaurus*, and two *Allosaurus* specimens, developed mounts that would last for decades.

problem with this approach is that the resulting estimate of weight depends entirely on the construction of the fleshed-out model. Because we cannot test how accurate these models are, our results are only very rough estimates.

The second approach relies on comparisons between fossil and living animals and is more empirical. The general idea is that there is a relationship between the width and length of limb bones and the weight of an adult animal. From thousands of measurements taken from all sorts of living animals, we can develop a general predictive mathematical model about these relationships. We can test the accuracy of this model by taking the same measurements from living animals whose weight is known. If the model performs well, we can take identical bone measurements on nonavian dinosaur fossils. Plugging the values into the model gives us extrapolated weights. Although perhaps more reliable than the first method, this is not a real test because no terrestrial animals that exist today even approach the size of the biggest dinosaurs. Hence we do not know how accurate the estimates from these models are at the extreme scale of the largest dinosaurs.

We might have confidence in the weights estimated for nonavian dinosaurs if different methods produced congruent (similar) results, no matter which method was used. Scientific investigators often employ congruence tests to evaluate whether the values they are getting are reasonable. Unfortunately, the weight values for nonavian dinosaurs estimated by these different methods are incongruent. For example, *Diplodocus carnegei* has been estimated to weigh 18.5, 11.7, or 5.8 metric tons, depending on the method used. These calculated estimates for the same animal vary by a factor of three!

So what is the biggest dinosaur? In terms of length we do not need experiments with models or sophisticated mathematics to figure this one out, just a very long tape measure. Even though only parts of skeletons have been excavated, *Argentinosaurus* and *Seismosaurus* are estimated to have been up to 50 meters long, making them the longest dinosaurs yet discovered. Although the weight of these dinosaurs has not been calculated, another sauropod, Apatosaurus (with a length of about 26 meters) has been estimated to weigh around 35 metric tons. *Argentinosaurus* and *Apatosaurus* were undoubtedly heavier, probably even heavier than the old heavyweight champion, *Brachiosaurus*. *Brachiosaurus* is not very long (only about 22 meters) but has a scale crunching weight of between 31.6 (bone measurement) and 87 (model dinosaur) tons. Considering that an adult male African elephant weighs 5.4 metric tons, these where truly earthshaking animals.

Question #15

What are the smallest dinosaurs?

FIGURE 19. The smallest dinosaur is the bee hummingbird, *Mellisuga helenae*, found only on Cuba. The male of this diminutive bird reaches an adult weight of only 1.95 grams.

The term "dinosaur" is often used synonomously with "giant," but some of the smallest terrestrial vertebrates are dinosaurs. Consider, for example, the bee hummingbird (*Mellisuga helenae*) from Cuba. The adult bird weighs only 1.95 grams and is small enough to fit into a teaspoon.

Although several ultrasmall nonavian dinosaurs have been collected, most are embryonic, hatchling, or juvenile individuals. The smallest nonavian adult dinosaur is the diminutive *Compsognathus longus* from Early Jurassic rocks of western Europe. Only two specimens of this animal have been discovered. One represents a small individual, about 70 centimeters long, including its long tail. It was found in the same lithographic limestones that preserved the specimens of *Archaeopteryx* in Bavaria. The other specimen is about twice as large and was found in similar deposits in southern France. Although longer because of the elongate tail, the body of *Compsognathus* was no bigger than a full-grown chicken weighing around 3 kilograms.

Several other dinosaurs, such as *Heterodontosaurus* from the Early Jurassic in South Africa, are almost as small as *Compsognathus*. Many dinosaurs have been found that are about the size of a small dog. One reason that small dinosaurs are hard to find is that their bones are very fragile and consequently are not frequently preserved. As our sampling and collecting methods become more sophisticated, more and more small dinosaurs are being collected. In the last ten years enough small dinosaurs have been collected to establish that dinosaurs evolved into a dazzling array of body sizes during the Mesozoic.

Question #16

What was dinosaur skin like?

The "skin" of reptiles is a complex system of scales separated by flexible joints. These joints are identical to the scales, except thinner, allowing flexibility. Our best clues about the skin of nonavian dinosaurs come from rare fossil occurrences and from birds, the closest living relatives of extinct dinosaurs. In general, the scales of birds are highly modified into feathers, except around the beak and on the feet. Because the foot structure in extinct theropod dinosaurs is very similar to that of living birds, they were probably covered with the same sorts of small scales found in living birds. The body coverings of dinosaurs more distantly related to birds are usually more difficult to determine. Nevertheless, a few spectacular fossils give us direct evidence about the body coverings of these animals.

FIGURE 20. Dinosaur skin is known from imprints, not from preserved skin. This imprint was associated with a *Corythosaurus* specimen and shows the nodular type of scales that are found on many dinosaur undersurfaces.

Records of soft parts, such as impressions of dinosaur skin, are preserved only under the most extraordinary circumstances, for example, in an environment that is very dry or that lacks enough oxygen to support decomposing organisms, such as the floor of a swamp or the deep sea. Burial must take place shortly after death, before decomposition begins. Only a few dinosaur specimens preserved under these special conditions have been discovered and in rare cases skin impressions are preserved.

The most spectacular of these specimens, and one of the greatest dinosaur fossils ever collected, is the American Museum of Natural History's *Edmontosaurus* mummy (*see* page 154). This specimen preserves impressions of skin over almost the entire body. The surface is covered by tubercles that vary in size. These tubercles can be categorized as large, diamond-shaped pavement tubercles and small, rounded tubercles. Around the base of the appendages, as well as at the neck and other joints, the skin is folded, like that surrounding the joints of an elephant, presumably to allow flexibility during movement.

Several other specimens have fossilized skin impressions. Many of these specimens were collected at what is now Dinosaur Provincial Park in Alberta,

Canada, the destination of many early expeditions. Here conditions appear to have been ideal for the preservation of soft part impressions. Most of these are very similar to the skin impressions of the *Edmontosaurus* mummy.

These specimens tell us that the general skin texture of dinosaurs was quite uniform. Impressions of Jurassic sauropod skin from Howe Quarry (*see "Diplodocus,"* page 100) and Late Cretaceous ceratopsian skin fragments (*see "Centrosaurus,"* page 168) both show the same general pattern of small tubercles, as do sauropod embryos from Argentina (see "*Sauropod embryos*," page 193). Other reports of feathred dinosaus (see "Liaoning" and "*Caudipteryx*," page 188) have the implications for the body covering, if not the skin, of many theropods.

Some dinosaurs had small bones called osteoderms embedded in their skin. These bones may have formed a sort of chain mail for protection. In some dinosaurs, especially ankylosaurs, these osteoderms were large bony plates arranged in a regular pattern. The bony plates of ankylosaurs were so extensive that they almost formed a shell like that of a tortoise, but the skin between the osteoderms allowed the shell to be flexible. In stegosaurs the osteoderms were shaped like small pebbles arranged almost randomly around the throat.

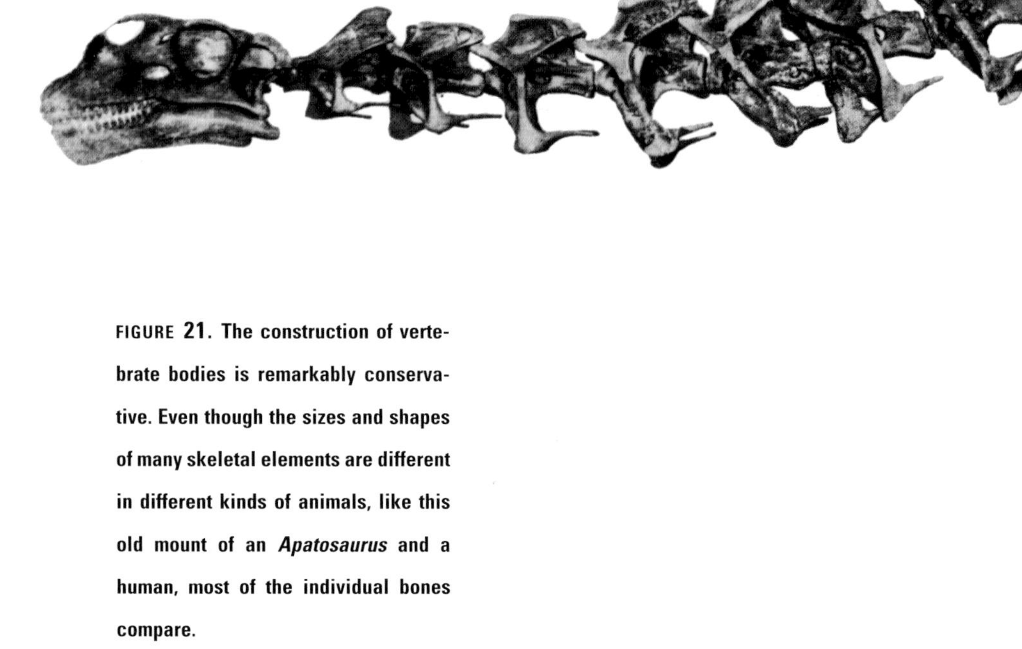

FIGURE 21. The construction of vertebrate bodies is remarkably conservative. Even though the sizes and shapes of many skeletal elements are different in different kinds of animals, like this old mount of an *Apatosaurus* and a human, most of the individual bones compare.

Question #17

How many bones do dinosaurs have?

Even though dinosaurs are the biggest animals ever to walk on land, they have about the same number of bones as other terrestrial animals with backbones have. The general body plan of all four-limbed animals (tetrapods) is conservative. In all these animals, whether bat, bird, or *Barosaurus*, the basic architecture of the skeleton is the same. Individual bones are usually so generalized that we can recognize them simply from their basic shape (morphology). Although distantly related, the femur (thigh bone) of an *Apatosaurus* is roughly the same shape as our own. The common morphology of these bones in all tetrapods is conclusive evidence that they all descended from a single unknown common ancestor that possessed a similar body plan.

The human skeleton has 206 bones. The smallest mammals (tiny shrews) and tiny reptiles (chameleons that can sit on the edge of a coin) have nearly the same number of bones as elephants have. Even the largest dinosaurs have the same number of bones, give or take a few. The only major differences are the additional vertebrae in the extremely long tails of some sauropods, such as *Diplodocus* and *Apatosaurus*. Since no skeleton preserving every bone of a large nonavian dinosaur has been excavated, however, we do not know exactly how many bones these animals had.

Question #18

What kind of teeth do dinosaurs have?

Dinosaurs belong to a group called the Archosauria, which includes crocodiles and pterosaurs (flying reptiles). One characteristic that all archosaurs inherited from their common ancestor is teeth that lie in sockets. These are unlike the teeth of many other reptiles, including those of lizards and snakes, which are simply attached to the inside of jaw. Fossils document that dinosaur teeth were replaced sequentially throughout life. As teeth wore out, new ones grew in from underneath in a specific pattern to take their place.

Among reptiles, dinosaurs show the greatest diversity of tooth types. Typically, carnivorous forms have sharp, serrated, daggerlike teeth. Primitive plant-eaters have leaf-shaped teeth, with small denticles that may have operated like small saws to shred vegetation. Many derived plant-eaters, such as the hadrosaurs (*see "Anatotitan,"* page 156) and ceratopsians, have thousands of teeth. These teeth form a large, renewable surface called a dental battery composed of the replacement teeth and the teeth currently in use. The large chewing surface and rapid tooth replacement are necessary for slicing or grinding tough, coarse plant material.

Not all dinosaurs have teeth. Teeth have been lost in the evolution of dinosaurs several times. Dinosaurs such as *Oviraptor, Struthiomimus,* and modern birds all lack teeth, but as anyone who has observed a hawk or a roadrunner knows, lack of teeth does not preclude an animal from being a skillful, predatory carnivore.

FIGURE 22. Pointed conical teeth, like those in *Allosaurus,* occur in all toothed theropods. Below right, a replacement tooth is growing from below and will push out the older tooth above it.

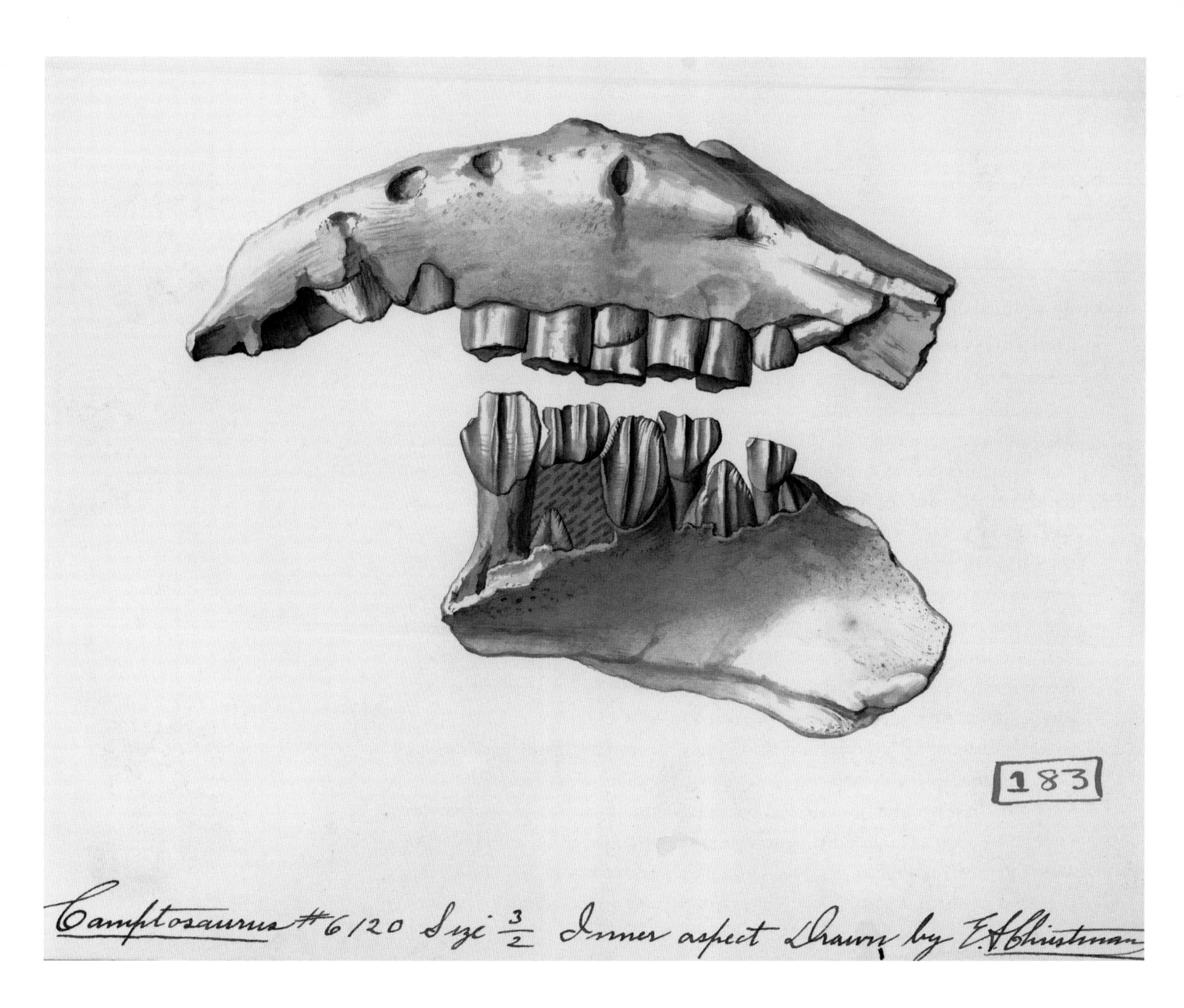

FIGURE 23. The teeth of the ornithischian *Camptosaurus* overlap and are shaped to fit together. The teeth always have a flat grinding surface, worn down by the teeth of the opposing jaw. These teeth are more advanced than those of the saurischians and gave rise to the large, complex dental batteries of the advanced hadrosaurs and ceratopsians.

Question #19

What color were dinosaurs?

Extinct dinosaurs are often portrayed as brightly colored as peacocks. These contemporary renderings are a stark contrast to the somber green or gray colors of dinosaur illustrations from the first half of the twentieth century. The earliest drawings and paintings of dinosaurs reflected the coloration of the living animals to which they were thought to be related, namely crocodiles, lizards, and other cold-blooded reptiles. The new interpretation is due in large part to changing perspectives about nonavian dinosaurs and their kin. The recent revitalization of the theory that birds are dinosaurs fostered a radical shift in the color palates used by dinosaur artists. Contrast the somber tones used by the legendary illustrator Charles Knight in the 1920s and 1930s with those of some of today's more outrageous illustrators. Or compare the coloration and pattern of dinosaurs in Steven Spielberg's 1993 movie *Jurassic Park* with that of Godzilla—a 1960s Japanese creation very loosely based on *Tyrannosaurus rex*.

Unfortunately, no direct fossil evidence is available to help us answer this question. No known dinosaur remains preserve skin coloration. Only under exceptional conditions does a fossil provide evidence of pattern or coloration. One such example is an extinct turtle collected in northern New Mexico's San Juan Basin. Because of unusual environmental conditions, this specimen preserves clear indications of the pattern of spots and speckles on the shell. The color of these spots and speckles, however, is not preserved. Although not impossible, finding an extinct dinosaur specimen that preserves skin color is highly improbable.

It seems logical to infer the color of dinosaurs from birds. Such inference is risky business. First, much of the coloration of birds comes from their plumage, and all available evidence suggests that most nonavian dinosaurs lacked a feathery covering. Nevertheless, even the exposed skin around the eye, on the bill, and on the feet of birds is often brightly colored and patterned.

Second, consider trying to determine what a mammoth looked like, using living elephants as a model. We would probably infer that these ice age elephants were largely hairless. A few of the frozen mammoths discovered in the Siberian and Alaskan tundra preserve long hair, indicating that our inference is incorrect. If our subject was horses, and all modern horse species were extinct except for the Mr. Ed type, guessing the color or pattern of a zebra from its skeleton alone would be inconceivable.

Most large animals display a color pattern. In the close relatives of the nonavian dinosaurs these color patterns are often associated with body characteristics, such as the delicate stripping on the muzzles of many crocodiles. Among living birds and crocodiles, some are strikingly colored and some are not. Given the incredible diversity of form in dinosaurs, some were probably patterned and colored. Unfortunately, we cannot distinguish the zebras and toucans of the group from those with the dreary color of a rhinoceros or a grouse.

Question #20

How fast did dinosaurs move?

Measuring the absolute speed of animals dead for over 65 million years is difficult. Nevertheless, some gross relative estimates are obvious from the morphology of dinosaur bodies. For instance, big graviportal sauropods with heavy club feet, while possibly capable of short bursts of speed, were probably not fast runners. Because of their immense size direct comparisons to living animals are difficult, but similarities in the shape and proportions of their limbs suggest that they were very similar to today's elephants. Conversely, long-legged ornithomimids ("ostrich" dinosaurs), were probably capable of running at sustained high speed.

FIGURE 24. Trackways of dinosaurs like this one from the Connecticut Valley are direct evidence of locomotion patterns and erect stance. Although some studies of trackways have yielded estimates of the speed of dinosaurs, there is no consensus.

Absolute speeds for extinct animals have been estimated by using complex models in which the length of stride (determined from footprint evidence) and leg length is compared to the same parameters in living animals. This comparison is difficult for several reasons, such as the difficulty of associating a particular set of fossil footprints with a particular species of dinosaur known only from a few skeletal remains. Complicating the situation is the fact that fossil bones are rare in rocks preserving footprints. The converse is also true, usually prohibiting precise identification of track makers.

Like many aspects of dinosaur behavior and biology, the speeds of dinosaurs have been subject to fantastic claims. Such claims are difficult to test, and the only available evidence (that determined from footprints combined with models based on living animals) suggests that, while larger dinosaurs may have been capable of running, they were predominantly slow-footed and unable to attain the speeds of even a moderately fast human runner.

Question #21

Did dinosaurs make sounds?

Among vertebrates, archosaurs probably have the greatest repertoire of voices and songs. Be it the bellowing of a lonely crocodile, the lyrical songs of a wren on a spring day, or the shrill call of a bird of prey, archosaurs have a tune that suits almost any taste. Because we cannot record the sounds of extinct, nonavian dinosaurs directly, we can only speculate about them by making inferences about potential sound-producing organs from parts that are preserved in the fossil record. This process is difficult and risky.

Much attention has surrounded dinosaur vocalization. Most of the scrutiny, critical or otherwise, has focused on the crested hadrosaurs, the lambeosaurines (*see* "A Phylogenetic Classification of Dinosaurs," page xv, and

FIGURE **25. This immaculately preserved skull of *Corythosaurus casuarius* from the Cretaceous Judith River Formation of Alberta reveals its internal structure in great detail. The expanded helmet formed by double pairs of nasal bones is hollow, leading to speculation that the expansion produced species-specific bellows and honks. As with other ideas about the function of the crests, this one is untestable and must remain unscientific speculation.**

"Corythosaurus," page 158). Many, like *Parasaurolophus* and *Corythosaurus*, had spectacular crests, protruding high above the skull roof. Soon after well-preserved fossil skull material was first collected, investigators noticed that the elaborate crests were actually hollow convolutions of the nasal passages. Eager to come up with an explanation, paleontologists proposed all sorts of wild speculations.

Because hadrosaurs were thought to be predominantly aquatic, one idea was that the crests worked like snorkels for breathing in deep water. This hypothesis was refuted both by the observation that the terminal ends of the crests lacked holes and by the fact that water pressure at a depth of 3 meters or so (the distance to the chest cavity from the top of the "snorkel") is so great it would prohibit the lungs from inflating. An alternative idea was that the crests provided air reserves for staying submerged, but unless the animals had some way of pressurizing the air, as in a scuba tank, the problem with water pressure prohibiting lung expansion made this theory impossible as well.

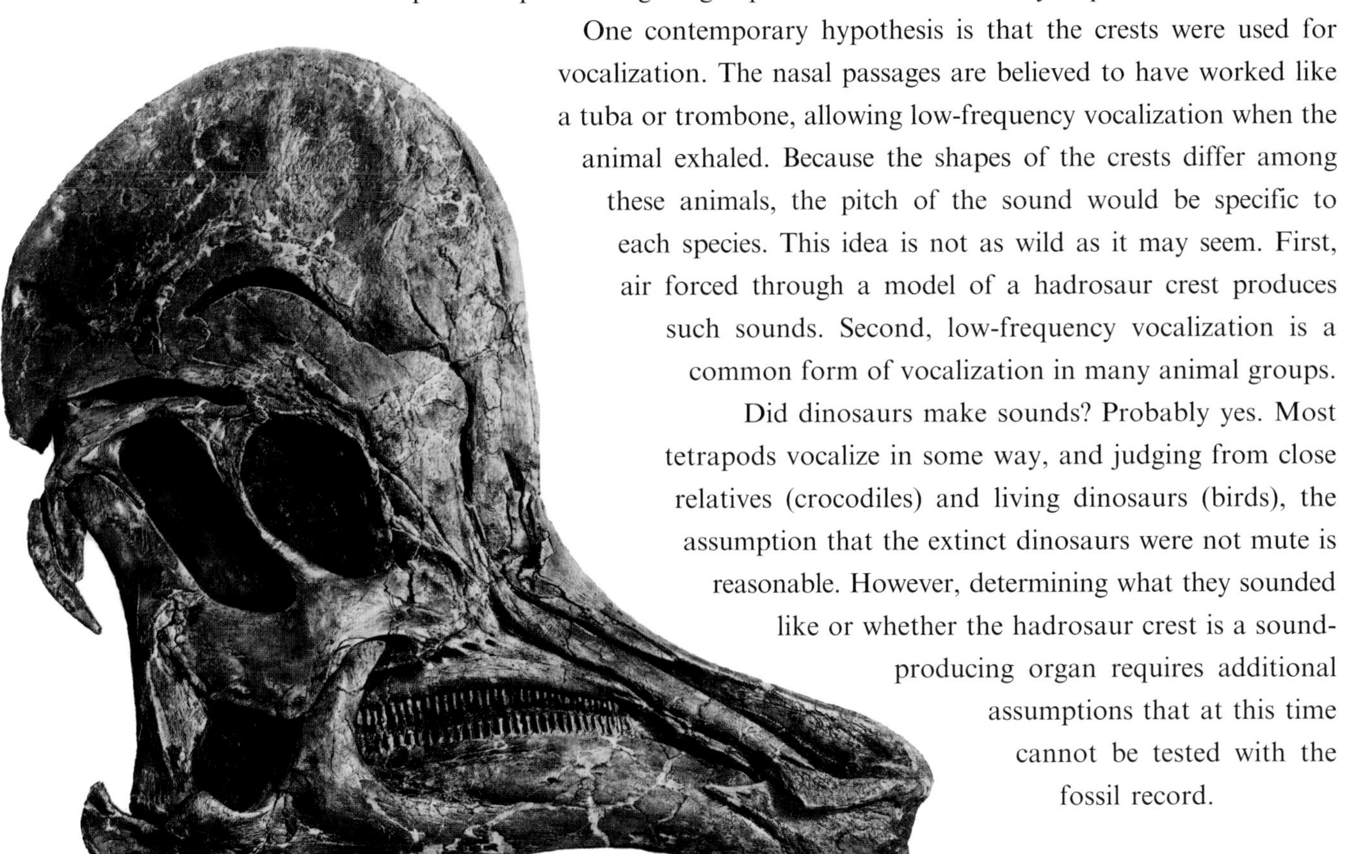

One contemporary hypothesis is that the crests were used for vocalization. The nasal passages are believed to have worked like a tuba or trombone, allowing low-frequency vocalization when the animal exhaled. Because the shapes of the crests differ among these animals, the pitch of the sound would be specific to each species. This idea is not as wild as it may seem. First, air forced through a model of a hadrosaur crest produces such sounds. Second, low-frequency vocalization is a common form of vocalization in many animal groups.

Did dinosaurs make sounds? Probably yes. Most tetrapods vocalize in some way, and judging from close relatives (crocodiles) and living dinosaurs (birds), the assumption that the extinct dinosaurs were not mute is reasonable. However, determining what they sounded like or whether the hadrosaur crest is a sound-producing organ requires additional assumptions that at this time cannot be tested with the fossil record.

Question #22

What were the primary sensory capabilities of dinosaurs?

FIGURE 26. ***Tyrannosaurus rex*** **has been interpreted as having stereoscopic vision, as primates do, in contrast to the largely lateral vision of birds and most tetrapods. As this well-preserved *Tyrannosaurus* skull (AMNH 5027) shows, however, the orbits require considerable reorientation to allow overlapping vision.**

The primary sensory capabilities of the nonavian dinosaurs are difficult to judge, but comparing a few skeletal characteristics with living animals allows some inferences.

It has been speculated that some extinct dinosaurs, especially theropods, had stereovision. Stereovision results when the eyes have overlapping fields of view and is important for accurately judging distances. When the eyes lie on the side of the head, as in a cow or a horse, visual fields do not overlap and vision is in two dimensions. In animals such as lions and humans, the eyes are directed forward, fields of view overlap, and vision is three-dimensional. Most dinosaur skulls are distorted during fossilization, so judging whether the eyes had overlapping fields of view is difficult, because the skulls are so difficult to reconstruct. It is at least plausible, however, that *Tyrannosaurus rex*, troödontids (*see "Saurornithoides,"* page 127), and some others possessed stereovision.

Visual acuity is more difficult to judge. In general, birds have keen eyes. Birds also see in color. We can only make inferences about the visual acuity of a nonavian dinosaur by examining different parts of its braincase and comparing them to corresponding regions that control visual acuity in living animals. Such study is fraught with problems. However, some loose generalizations are possible. The part of the brain that processes visual information is the cerebrum. Based on comparison of braincase morphology, cerebrum size varies among dinosaurs and is developed to its highest degree in the maniraptors. Not surprisingly, the maniraptors include the visually acute group birds, so we have a phylogenetic basis to surmise that the visual capabilities of the extinct nonavian maniraptors were not radically different from those of their modern relatives.

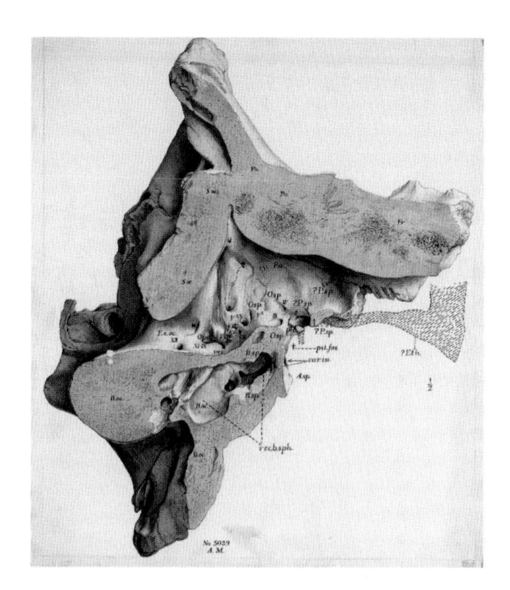

FIGURE 27. The braincase of *Tyrannosaurus rex* (AMNH 5029) shows the size of the cavity that held the brain. The actual brain was probably smaller. Studies of braincases show that most dinosaur brains were quite similar in relative size to the brains of living reptiles such as turtles, lizards, and crocodiles.

Like vision, scent capability is hard to judge from fossil remains, and again the best evidence comes from the inferred morphology of the brain. At the very front of the brain all vertebrates have elongate protrusions called olfactory bulbs, where chemosensory information is processed. In general, the olfactory bulbs of dinosaurs are small, suggesting a poor sense of smell. Elaboration of the sinus passages in some dinosaurs (such as ankylosaurs and lambeosaurines), may reflect a more developed sense of smell, but there are alternative explanations concerning the function of these passages (*see* "Did dinosaurs make sounds?," page 34).

Acoustic capabilities of nonavian dinosaurs are probably the most difficult of the primary sensory capabilities to examine. Hearing is based largely on development of the inner ear, and very little is known about the anatomy of this region in nonavian dinosaurs. Because birds have an inner ear that is advanced over that of crocodiles, at least some nonavian dinosaurs probably also had an advanced inner ear. The question of which nonavian dinosaurs had birdlike ears and which ones were more like crocodiles is currently being evaluated by specialists at the Museum using CAT scan data (*see* "How are fossils prepared?," page 87).

Question #23

How intelligent were dinosaurs?

There is no direct way to measure intelligence in a fossil animal. Even in living animals this quantity is difficult to measure. Consider the social minefield of trying to measure this in our own species. Nevertheless, many scientists have tried to measure relative intelligence in dinosaurs.

The most common approach has been to calculate what is called an encephalization quotient. This measure compares the size of the brain to the size of the body. Calculating such values is not a simple proposition in fossil organisms, where estimates of body size can easily be off by 50 percent or more (*see* "How large were the biggest dinosaurs?," page 22). Furthermore, brain size (measured by the volume of the braincase) is often determined from very fragmentary or distorted fossil skulls through divination. This problem is compounded by the fact that in most reptiles, except birds and most other maniraptors, the brain does not fill the cranial cavity. Thus, the volume of the braincase is not an accurate measure of brain size.

We can roughly estimate brain size, however, by measuring braincase volume. For example, we can determine that some dinosaurs had a larger brain (a higher encephalization quotient) than others and that dinosaurs in general did not have inordinately small brains when compared with those of crocodilians, turtles, and lizards. Troödontids (*see "Saurornithoides,"* page 127) had the largest brain of any nonavian dinosaur. Translating this directly to intelligence is risky, however. The closest immediate relative to humans, Neanderthals, had a bigger brain than we do, but few would argue that these cave people with sloping foreheads were more intelligent than we.

FIGURE 28. Although most dinosaurs had brains similar in relative size and complexity to those of living crocodiles and lizards, some of the birdlike theropods had much larger brains. Some of these brains were close in size to those of smaller-brained mammals and birds. This braincase of an advanced theropod, the troödontid *Stenonychosaurus*, which is closely related to birds, shows a large brain cavity, comparable in relative size to the brain of a bird.

Better clues to intelligence are reflected in brain organization rather than absolute size. The brains of all vertebrates have three main parts: the brain stem, the cerebellum, and the cerebrum. The brain stem is closest to the spinal cord and is involved in regulating heartbeat, respiration, and other vital functions. The cerebellum is tied to muscular coordination, and the cerebrum is associated with several complex sensory functions, with memory, and with motor control. Derived or advanced evolutionary modifications in these regions involve enlargement, subdivision, and complexification. In most vertebrates, the size of the brain stem is fairly uniform. The cerebrum and cerebellum in both birds and mammals, however, is extremely large, with a convoluted surface. Enlargement and modification of the cerebrum is associated with enhanced sensory and memory capabilities that might be termed intelligence.

Fossil dinosaur specimens with well preserved braincases can give clues about brain organization and evolution. In some animals, such as troödontids and dromaeosaurs, the cerebrum and the cerebellum appear large and birdlike. So how smart were dinosaurs? Some were probably about as intelligent as primitive modern avians, like rheas or ostriches, but not as intelligent as some more advanced birds, in which the brain is even larger and more complex.

Question #24

What did dinosaurs eat?

FIGURE 29. Coprolites such as this 135-million-year-old specimen from the Morrison Formation can yield useful clues regarding the diets of dinosaurs. Unfortunately, connecting specific coprolites to specific dinosaurs is nearly impossible, so we still do not know what particular dinosaurs ate, except in very general terms.

Dinosaurs are diverse. Animals of almost every size and shape imaginable have existed during the history of the group. Dinosaur diets were probably also diverse. Determining what extinct dinosaurs ate is difficult, but we can infer some aspects of their dietary preferences. Traditionally this information has been derived from direct evidence, such as stomach contents, and indirect evidence, such as establishing a correlation between particular body characteristics and diets of living animals and then inferring habits for dinosaurs.

Animals such as house cats and dogs have large, stabbing canine teeth at the front of the mouth and smaller, equally sharp teeth farther back in their jaws. Many of these animals are also armed with sharp claws. The advantage of teeth and claws as predatory tools is obvious. Now consider animals like cows, horses, rabbits, and mice. These animals have teeth at the back of the jaw that are flat, analogous to, and having the same function as a grindstone. Unlike the meat-slicing and stabbing teeth of carnivores, the teeth of these animals grind and shred plant material before digestion.

More clues exist in other parts of the skull. On a carnivore like a cat or a dog the jaw joint is at the same level as the tooth row. This gives these animals a mechanical advantage for closing the jaws with tremendous speed and forces the upper teeth to occlude against the lower teeth with great precision. In herbivorous animals rapid jaw closure is less important. Because the flat teeth of herbivores work like grindstones, however, the jaws must move both side-to-side and front-to-back. The jaw joints of many advanced herbivores, such as cows, lie at a different level than the tooth row, allowing transverse tearing, shredding, and compression of plant material. If we extend such observations to extinct dinosaurs, we can infer dietary preferences (such as carnivory and herbivory), even though we cannot determine the exact diet. The hadrosaurs are a good example of a group whose jaw joint is below the level of the tooth row, which probably helped them grind up tough, fibrous vegetation.

Question #25

Did dinosaurs fight?

Most vertebrates engage in combat—even moronic, pastoral herbivores like cows and sheep. Usually fighting is initiated through competition for territory, mates, or food. As with other types of behavior, direct evidence in the fossil record is rare. Occasionally we are lucky. In a few cases, we have extraordinary fossils preserved that demonstrate fighting between members of the same species and different species.

FIGURE 32. Direct evidence of predation or other violent interactions between dinosaurs is rare. This dromaeosaur skull has a wound apparently caused by the bite of a carnivorous dinosaur of similar size.

In interpreting this evidence an important distinction must be made between hunting and fighting behavior. The first involves fighting between predators and prey. The second occurs among or between species but—usually not for the intended purpose of one animal killing the other and eating it.

Hunting occurs all the time, and evidence of carnivory is common in fossils. One famous example is the partial skeleton of *Apatosaurus* from Como Bluff that is mounted at the base of the American Museum of Natural History's *Allosaurus* mount (*see* page 110). Some of these vertebrae have deep grooves thought to be the tooth marks of a large *Allosaurus*-like carnivore. Whether this find represents a kill or is an indication of scavenging cannot be determined; it is not unlikely, however, that large carnivorous dinosaurs like *Allosaurus* preyed on animals larger than themselves.

The most spectacular example of dinosaur combat was discovered in Mongolia. The fossil documents a carnivorous theropod, *Velociraptor*, entwined in a death struggle with a herbivorous *Protoceratops*. As difficult as it is to believe, these specimens appear to be preserved in the heat of battle: The right arm of *Velociraptor* is caught in the mouth of *Protoceratops* with the hand of *Velociraptor* grasping the face of the herbivore. And the large, powerful, slashing foot claw of *Velociraptor* is embedded in the neck of *Protoceratops* near where the carotid artery would lie in life. The exact circumstances surrounding the scene are unknown and hard to imagine. It has been suggested that these specimens represent a chance accumulation. The placement of the

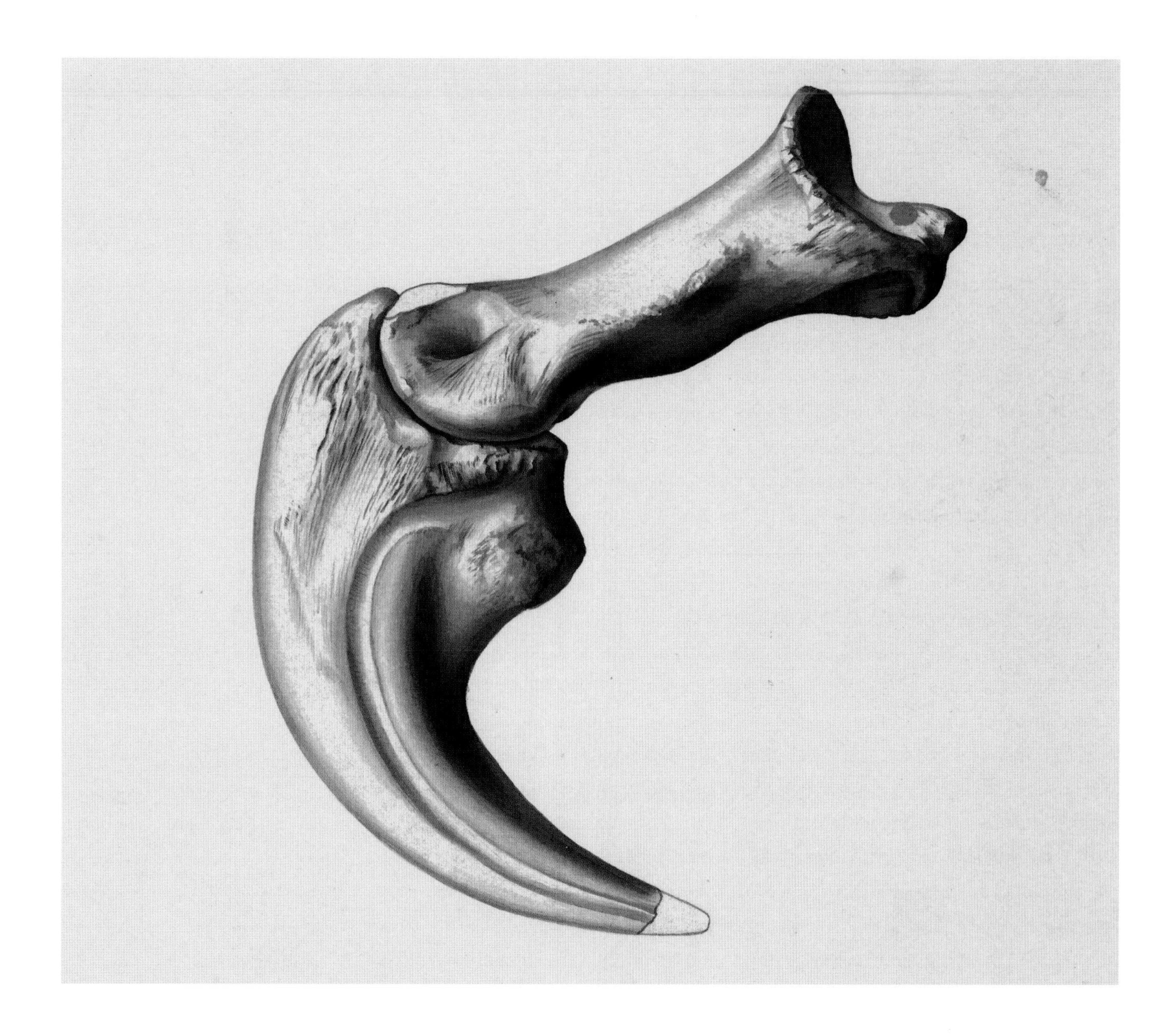

FIGURE 33. The claw of *Allosaurus* (AMNH 5753) strongly suggests the function of catching and handling prey.

Velociraptor arm, however, at least indicates that the two animals were associated during life.

Although some specimens have been recovered, examples of fighting between members of the same species are more rare. Fossil specimens showing indications of combat are predominantly theropods, not surprising, considering their probable predatory nature and ferocity. A spectacular example is a large *Tyrannosaurus* collected in South Dakota. This specimen is really beaten up, so much so that one eye is nearly closed by abnormal healing of the surrounding bone. On another part of the skull, the broken tip of a tooth from another tyrannosaur is embedded. Around this tooth are the characteristic marks of bone growth that occurs during healing, indicating that this animal survived the attack. Many carnosaurs, including the American Museum of Natural History's *Tyrannosaurus rex*, show signs of healed injuries that may have resulted from combat.

A new example of fighting between members of the same species was collected in 1991 from the same locality that produced the fighting *Velociraptor* and *Protoceratops*. This evidence is on another *Velociraptor* skull. The frontal bones of this specimen show two parallel rows of small punctures that appear to represent tooth marks. The pattern of tooth marks corresponds identically to the spacing of teeth in the upper jaw of *Velociraptor*. The proximity of the tooth marks to the brain indicate that these are the marks of an attacker's lethal blow. Healing did not occur, and the specimen was not extensively scavenged, suggesting that this animal was killed in a fight with another of its own kind, or at least another species with similarly sized jaws.

What about combat involving plant-eating dinosaurs? No direct evidence conclusively demonstrates battles between nontheropods. Many modern herbivores (for example, bison, deer, mountain sheep, and wild horses) often fight for territory and mates. Although they usually do not continue to the death, these fights are ferocious, and their violence would be magnified several times if they were between large dinosaurs weighing several tons.

Several indirect clues have been used to infer that ritualistic fighting occurred between some ornithischian dinosaurs, and an extensive body of literature has been developed on the subject. Most of these arguments focus on the correspondence between structures used in fighting of animals today and those typically found in dinosaurs. One of the most popular scenarios is that pachycephalosaurs used their dome heads as battering rams like bighorn sheep, allowing the top of the skull, with its 25-centimeter-thick crown, to absorb the force of the momentous collision. Such proposals are reasonable, but reason alone does not always make for an empirically testable scientific hypothesis. Similarity in form between a modern animal with a known behavior and a fossil animal whose behavior cannot be known can be deceiving. The possibility always exists that alternatives (such as sexual display) account for these spectacular features.

Question #26

Did dinosaurs travel in herds?

Figure 4 (see page 1). R. T. Bird unearthing a huge slab of Early Jurassic theropod footprints near Tuba City, Arizona in 1937.

Whether or not nonavian dinosaurs traveled in herds is a loaded question. If we hypothesize that they traveled in herds we are inferring that they were active, intelligent animals with a highly developed social structure. The debate about whether nonavian dinosaurs herded has been incorporated into popular illustrations of these animals over the last several decades. For instance, compare restorations of dinosaur communities done in the first half of this century, which portray dinosaurs as dim-witted loners, with modern illustrations depicting them in herds of plant-eaters pursued by packs of carnivores. Aside from the obvious present-day evidence of flocking birds, what is the evidence suggesting that extinct dinosaurs moved in herds?

Most of the evidence is highly conjectural. Often cited are mass deposits of bones, as well as evidence from trackways. There are several alternative explanations for massive bone beds; these are explored elsewhere in this book (*see*, for example, *"Coelophysis,"* page 107, and *"Diplodocus,"* page 100).

Some of the most definitive evidence about dinosaur behavior comes from studying sequences of fossil footprints called trackways. Over the last decade there has been a resurgence in interest in dinosaur tracking. Several track sites provide evidence that suggests herding behavior in several groups of dinosaurs. Among these is the famous Davenport Ranch site in Texas, first discovered in the early 1940s by R. T. Bird, a fossil collector for the American Museum of Natural History. Detailed analysis of this site has shown that this trackway represented a herd of twenty-three sauropods traveling over muddy ground at a pace of about 2 meters per second (*see* "How fast did dinosaurs move?," page 33). The pattern of overlapping tracks revealed that the largest animals were at the head of the herd, followed by smaller animals. In a significant number of instances, animals followed in line. Other trackways, in the Purgatory River Basin of Colorado and in Korea, show herds of large sauropods, not traveling in a cluster, but moving side by side as a horizontal

front. At one spectacular locality in the Purgatory River Basin the animals' paths undulate from right to left in concert.

Analysis of the spacing of different individuals, or "herd structure," especially relating to sauropods, has received intense attention from paleontologists. Some paleontologists have even speculated that the herds were like elephant herds, in which smaller juveniles traveled in the core of the herd surrounded by large adults. Although the Davenport Ranch track site represents individuals of many different sizes, no unequivocal evidence in the tracks documents a particular pattern of size structuring in the herd beyond the fact that the larger animals were at the front. Other track sites show various types of size structuring, but this variation may be just a random occurrence based on a small sample size.

FIGURE 34. The only direct evidence of herding behavior among extinct dinosaurs comes from trackways. This spectacular dinosaur stomping ground, called the Davenport Ranch trackway, has been studied extensively. Analysis has indicated that a number of large sauropods were moving as a group. The pattern of overlap of the tracks indicates that the largest animals were at the head of the pack.

Theropods, which are less common than herbivores (*see "Allosaurus,"* page 110, and "Were dinosaurs warm-blooded?," page 52), are rarely represented in trackways. One occurrence is Bird's famed Paluxy River sites, where some evidence for a pack of theropods has been preserved. Part of this trackway now lies behind the *Apatosaurus* mount at the American Museum of Natural History in the new Hall of Saurischian Dinosaurs. We will examine this trackway later (*see* "The Paluxy River Trackway," page 181), but we should point out that the extent of the trackway was much larger than the area excavated.

The best evidence that theropods traveled in packs is a giant trackway from the Late Cretaceous of Bolivia known as Toro-Toro. Dinosaur tracker Giusseppe Leonardi describes the scene: *"In another rocky pavement, about 300m downstream, are eight parallel trackways of large sauropods (six adults, two juveniles) which advanced along the shore in a front…both juveniles moved together next to an adult. They were followed, undoubtedly shortly afterwards, by a pack of at least 32 medium sized carnosaurs, which sometimes stepped into the sauropod footprints and were moving in the same general direction and on the same front."*

Question #27

How did dinosaurs mate?

The sex life of dinosaurs has always been a subject of lively discussion. Mating behaviors among terrestrial vertebrates are generally conservative, and we have no reason to believe that the extinct dinosaurs strayed from the normal pattern. Usually the male mounts the female from behind, and sperm is passed to the female. In many animals sperm is transferred through an intromittent organ, such as the penis in mammals or the paired hemipenises in snakes. Living crocodilians have a rudimentary intromittent organ; most birds lack intromittent organs, except for primitive living birds, such as ratites (ostriches, kiwis, emus, and so on) and ducks and geese, which have a small penis. But what about nonavian dinosaurs? Some evidence is provided by analyzing the question phylogenetically. Because both crocodiles and primitive birds have intromittent organs, nonavian dinosaurs probably also inherited this feature from their common archosaurian ancestor. Additional evidence comes from recent study of the pelvis and anterior tail vertebrae in dinosaurs. Differences in these areas may allow sexes to be differentiated. In specimens thought to be males there are accessory areas for muscle attachment that could have formed the attachment points for muscles involved in penile function.

The physical act of mating in living crocodiles is often complex, with the long tails and awkward bodies making the process challenging. The mating of nonavian dinosaurs must have been a spectacle, and the ground must have shook as a result of the amorous affections of sauropods or large theropods.

In 1991, when the American Museum of Natural History christened its *Barosaurus* mount in the Roosevelt Memorial Hall (*see "Barosaurus,"* page 105), the choice of the pose was greeted with some disbelief by many of our colleagues in the scientific community. What they objected to was the possibility that this giant sauropod could rear up on two legs, supported only by its hindquarters. Although we have no direct evidence about the behavior of these animals except what has been preserved in fossil trackways (*see* "Did dinosaurs travel in herds?," page 45), it seems obvious to us that they could. If not, how else would they have mated?

FIGURE 35. Dinosaur eggs from the Cretaceous of Mongolia that were found in the 1920s by American Museum of Natural History expeditions were the first well-preserved dinosaur eggs and nests ever found.

Question #28

How did dinosaurs give birth?

Although the mode by which some nonavian dinosaurs gave birth is still uncertain, all dinosaurs probably laid eggs. Two sources of information support this proposal. First, no archosaurs have been conclusively demonstrated to give birth to live young, and all living archosaurs (birds and crocodilians) lay eggs. Without contradicting evidence, there is no reason to believe that any of the extinct archosaurs (including nonavian dinosaurs) strayed from this general pattern. Second, a pervasive piece of evidence is that eggs have been collected for several different groups of dinosaurs. Because many have been found in nests and a few even contain embryos, paleontologists can identify to which type of dinosaur many of these eggs belonged.

Some paleontologists have suggested that a few dinosaurs gave birth to live young. Live birth has evolved independently in many different sorts of animals. Consider guppies, mammals, and some snakes and lizards. Fossil evidence exists that some marine reptiles (ichthyosaurs) gave birth to live young, because some specimens apparently died while giving birth. There is no direct evidence, however, to support the idea that any dinosaur gave birth to live young.

Question #29

Did dinosaurs care for their young?

FIGURE 36. The discovery of Cretaceous dinosaur eggs and nests in Mongolia provided evidence for the reproductive habits of some dinosaurs, but more detailed ideas about dinosaur behavior have been difficult to test objectively.

From fossil evidence alone this question is very difficult to answer. Because behaviors are not preserved in the fossil record, we can only make inferences from indirect evidence. Parental care can be divided into two types of behavior: prehatching (building nests and incubating eggs), and post-hatching (feeding the young and guarding the nests).

Most of our direct evidence comes from alleged dinosaur rookeries. Several have been excavated in eastern Montana, where a large concentration of dinosaur nests was found at a place now called Egg Mountain. Most of these probably belonged to the hadrosaur *Maiasaura*. Preserved in these nests are the bones of baby dinosaurs. The finds at Egg Mountain and other sites around the world document that dinosaurs laid their eggs in nests.

The nests at Egg Mountain are reported to be equally spaced, separated by a space corresponding to the length of an adult *Maiasaura*. From this arrangement scientists have inferred that the nests were separated in this way to allow incubation in a tightly packed nesting colony. Although this interpretation is open to challenge, the discovery of *Oviraptor* adults on top of *Oviraptor* egg clutches (as determined by embryos in some eggs), is relatively powerful evidence that at least these dinosaurs incubated their nests.

Evidence for parental care following hatching is much more controversial. Behavioral speculation based on indirect fossil evidence is dangerous because the data is not always as unambiguous as might appear. At Egg Mountain, many nests contain baby dinosaur bones. Not all the dinosaurs in the nests are the same size. Many of the small bones found in the nests are associated with jaws and teeth, teeth that show signs of wear. It seems reasonable to assume that the wear was caused by the chewing of the coarse plants that were the hatchlings' diet. Because the young were still in the nest, this food may have been brought to the rookery by foraging adults. This line of reasoning suggests that these animals had an advanced system of parental care. A closer look at the evidence clouds this interpretation. Analysis of dinosaur embryos indicates that worn surfaces are present on the teeth of juveniles even before hatching. Just as a human baby moves inside the mother before birth, baby archosaurs inside the egg grind their teeth, wearing the surfaces in some spots. Thus, the fossil evidence for an advanced parental care system in extinct dinosaurs is suggestive but inconclusive, and it is hard even to imagine the sort of paleontologic discovery that could settle this debate for good.

The strongest evidence that extinct dinosaurs had some form of advanced parental care system is based on an understanding of the phylogenetic relationships among dinosaurs and their closest living relatives. Living dinosaurs, even primitive ones such as ostriches and kiwis, exhibit parental care, so some form of parental care can be inferred to have existed in the last common ancestor of all kinds. Although unappreciated, crocodiles are also caring parents. They build nests, guard the nests, and in some cases dig their young out of the nest when they hear the chirping young ones hatching. The young even communicate with each other while still in the egg by high-frequency squeaks (as birds do). Some evidence suggests that this squeaking is a cue for the synchronization of hatching. Since birds and crocodiles share a common ancestor, the simplest explanation for the characteristics they share (such as nest building and some form of parental care) is that they evolved only once—that these attributes were present in their common ancestor and passed on to its descendants. Because extinct dinosaurs also descended from that ancestor, the simplest and most general theory is that extinct dinosaurs also shared these characteristics, even though they cannot be directly observed and we cannot be sure how elaborate their parental care was.

Question #30

Were dinosaurs warm-blooded?

All living things depend on chemical reactions in the body to sustain life, such as the burning of sugar to produce energy while releasing carbon dioxide and water. These reactions are called metabolism. Their rate depends on internal body temperature. In general, each chemical reaction runs most efficiently at a particular optimum temperature.

Some animals, such as living mammals, have internal mechanisms that regulate body temperature to keep it near the optimum level. Because optimum temperatures for chemical processes are usually higher than that of the outside environment, these animals are commonly called warm-blooded, or endothermic. In other words they create their own metabolic heat internally to maintain a nearly constant body temperature.

Other animals, such as crocodiles, lizards, and turtles, vary their internal temperature depending on their activity level. These animals maintain their internal temperature within optimal limits by taking advantage of external factors through their behavior, such as lying in the sun to warm up or in the shade to cool off. They are commonly called cold-blooded, but this term is not accurate. It is more accurate to say that animals such as lizards have a variable body temperature because "cold-blooded" animals such as desert lizards often have body temperatures much higher than those of any "warm-blooded" animal. These animals develop their heat from external environmental sources rather than metabolic sources. Therefore, "ectothermic" is a better term than "cold-blooded."

Were nonavian dinosaurs more like lizards or more like mammals in their internal temperature regulation? Evaluating the mechanisms of temperature control in a living animal is often difficult, and the question cannot be answered positively for the extinct dinosaurs. Some dinosaurs (birds) were and are definitely warm-blooded. Birds have a high activity level, along with a steady and internally regulated temperature. At what point dinosaurs evolved

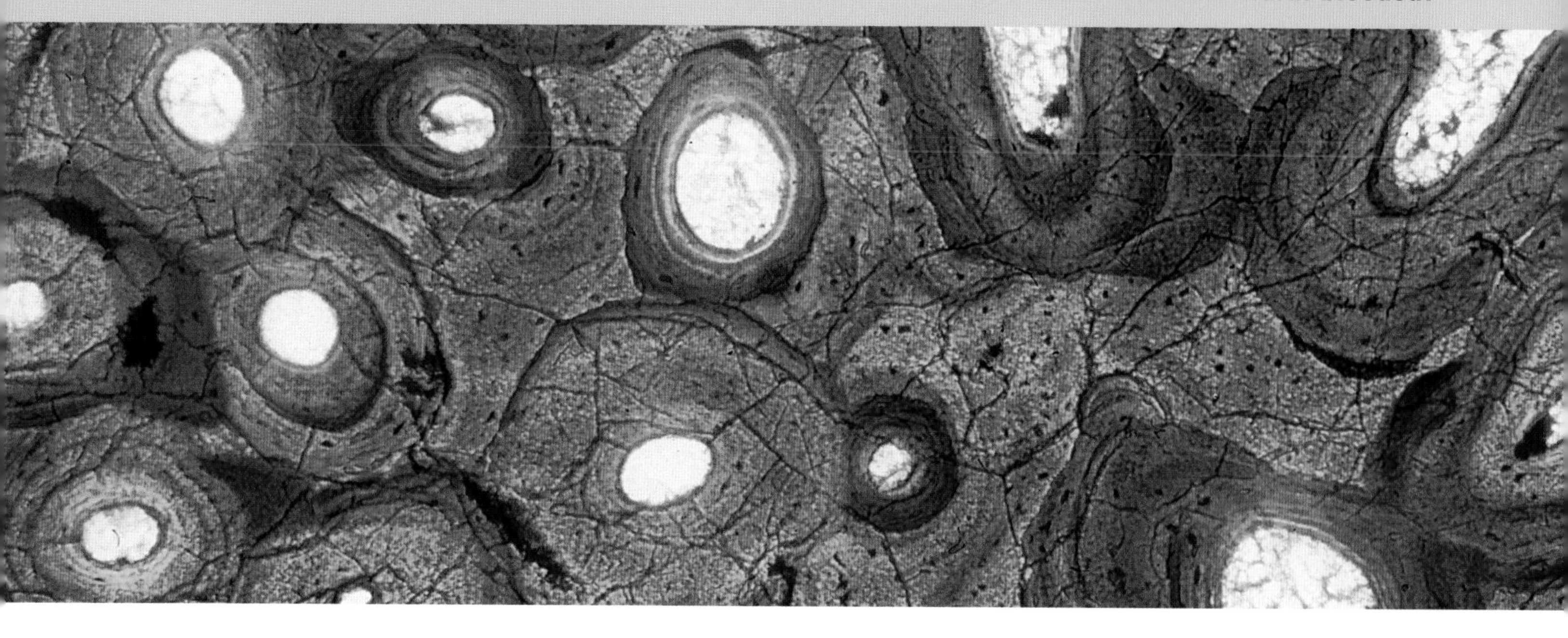

FIGURE 37. Thin sections of fossil bones examined microscopically give important clues regarding the microstructure of fossil bones, which may provide evidence for dinosaur physiological properties. In addition, they demonstrate how bones can be fossilized with almost exact fidelity. This thin section is from a leg bone of *Allosaurus*.

endothermy, however, is a difficult question. Indirect evidence used to infer the metabolic pattern of extinct dinosaurs comes from several sources, among them analysis of the fine structure and composition of bone, and inferences based on behavior and proportions of different kinds of animals coexisting in a habit (what ecologists call community structure).

Bone looks like a solid mass of nonliving material. Examining a thin section of bone under a microscope, however, shows that this is not the case. Many structures are visible. Holes carry small blood vessels or nerves; cavities form where bone has been resorbed; and open areas occur where cells secreting bone live or have lived. Bone microstructure is so complex that many things about animals (including clues about phylogenetic relationships) can be determined from it. In some ectothermic animals concentric rings, thought to indicate annual growth (as tree rings do), are common. Endothermic animals have a complex system of closely spaced cavities caused by the resorption and redeposition of bone called the Haversian system. The exact function of this system is unknown. One idea is that the Haversian system permits bones to be remodeled rapidly. This ability may allow endothermic animals to change quickly in size and shape during growth.

One way to test for endothermy is to cut dinosaur bones into very thin slices and compare them under a microscope with the bones of living animals whose temperature control systems are empirically understood. Although this procedure has been done for a variety of nonavian dinosaurs, the jury is still

FIGURE 38. Generally, specimens of carnivorous dinosaurs like this tröodontid are very rare. However, at some localities, they are very common.

out. In most nonavian dinosaurs, the microstructure of bone appears more like the bone of endothermic animals than of ectothermic animals, but this evidence is not conclusive. To standardize the observations, we must compare animals of similar size. Unfortunately, no nonavian, dinosaur-sized, ectothermic animals are alive today, and we have only begun to study small dinosaurs. Recent analysis of the bone fine structure of a primitive bird from the Cretaceous of Argentina suggests that this animal may have had a metabolism more like ectothermic modern reptiles than like endothermic modern birds.

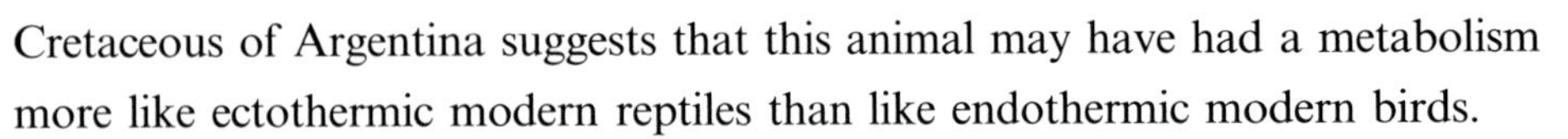

Other indirect evidence comes from the inferred behavior and community structure of dinosaurs. Most of the arguments stem from over interpretation of the available data. An example is the conception of dinosaurs as highly active animals, similar in behavior to modern mammals. Such behavior requires a high metabolic rate, such as that found in endotherms. Many structures observed in dinosaurs, such as the long, ostrichlike legs of ornithomimids, suggest active behaviors, but real evidence is lacking and is hard to generate from traces and remains of animals extinct for at least 65 million years. Instead, most of this "evidence" consists of subjective and unscientific claims about how these animals behaved or regulated their metabolism.

The relative number of predators to prey in a community may provide clues about an animal's metabolism. In modern ecosystems predators with a constant high metabolic rate (endotherms such as lions) need to eat much more than predators with lower rates (ectotherms such as Komodo dragons). Proportionally, there are many more prey in communities dominated by endothermic predators than in ecosystems dominated by ectothermic predators. Consequently, there are proportionally far fewer carnivores in endothermic communities than there are herbivores (their prey). Some dinosaur paleontologists have attempted to apply such a reasoning to the fossil record by contrasting the numbers of individuals of fossil carnivores and herbivores found at a locality. Their findings indicate that in many nonavian dinosaur communities presumed carnivores make up a tiny percentage of the total number of animals, suggesting that dinosaurs were warm-blooded.

There are serious problems with this argument. First, some localities don't empirically conform to this model. In many communities, such as the Late Cretaceous communities from the Gobi Desert, carnivorous dinosaurs are exceedingly common, to the point of being the most common dinosaur remains collected. The real difficulty with the predator-to-prey ratio argument is that it requires the fossil record to preserve relative numbers of kinds of animals with complete efficacy. The record can be demonstrated, however, to be a very poor indicator of these quantities. For example, although fossils of juvenile and small dinosaurs are relatively rare, both were almost certainly present in high numbers. The ratios of different types of dinosaurs probably have more to do with the vagaries of fossilization and the particular talents and tastes of fossil collectors than with how many predator and prey animals really existed.

Were nonavian dinosaurs warm-blooded? The evidence is still equivocal, and most claims that all dinosaurs are "warm-blooded" are speculative. There is no clear-cut evidence that dinosaurs were either cold-blooded or warm-blooded, except that dinosaurs evolved endothermy sometime in their history, as documented by living birds.

Question #31

How fast did dinosaurs grow?

Dinosaurs are the largest animals ever to walk on land. How did they get so big? Rates of dinosaur growth can be explored best by looking at living animals, which vary considerably in growth rate. In crocodiles (and in other cold-blooded animals), high growth rates are correlated with increased food. Even at the relatively high growth rates of well-fed crocodiles, it would take decades for a large sauropod to grow from hatchling to adult. This fact has led to speculation that nonavian dinosaur growth rates were higher than those of nondinosaurian reptiles, perhaps approaching the growth rates of birds. Bone microstructure seems to support this view because dinosaurs have a particular type of bone that has been putatively correlated with fast growth rates. However, all of this evidence is circumstantial, and needs to be evaluated empirically.

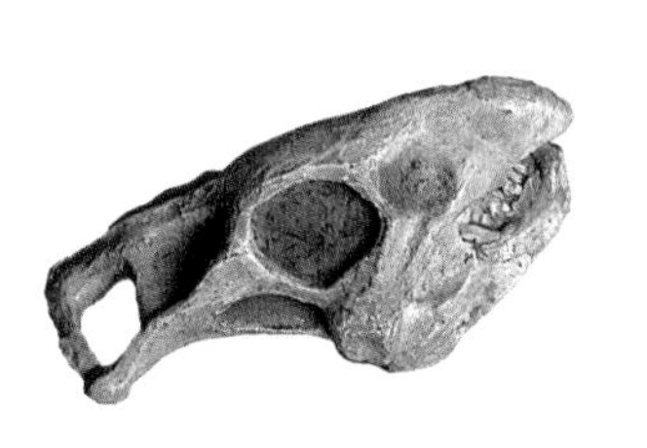

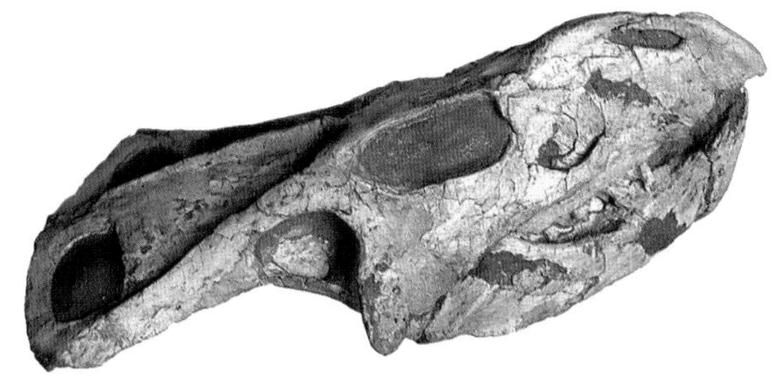

By comparing adult body weight to growth rate in living animals, we can predict dinosaur growth rates. A model is developed based on values measured in a wide variety of animals of different sizes and metabolisms. Then we can simply plug the estimated body weight of adult nonavian dinosaurs into the model and determine the growth rate.

This method sounds good, but just as in determining dinosaur weights, using a model poses some problems. First, estimating the weight of extinct dinosaurs is difficult (*see* "How large were the biggest dinosaurs," page 22). Second, because, like us, dinosaurs have what is called determinate growth (meaning that they reach a maximum size and do not keep growing throughout life), the growth rate varies through life, slowing to a near stop at maturity. Third, the metabolism cannot be considered to be just like that of mammals, birds, or crocodiles. Any estimate of growth in these animals must accommodate the huge variation resulting from "slop" in these values and thus is likely to be uninformative, or even misleading.

Despite drawbacks, models provide our best estimates. In 1978 Ted Case, an ecologist from the University of California at San Diego examined the problem. He included data on *Protoceratops* and the sauropod *Hypselosaurus*, both of which have been tied to egg fossils and are known from material extensive enough to enable an estimate of weight. Case found that, if *Protoceratops* grew at the slowest rate found in extant reptiles, it would reach top body size in 33 years (with a variance between 26 and 38 years), but because sexual maturity occurs at a much smaller size than top body weight, Case predicted that after 8 to 16 years *Protoceratops* would be part of the breeding population. If growth rates matched top ones in living reptiles, maximum size would be achieved after 12 to 23 years. For the much larger *Hypselosaurus*, the range of years to reach maximum size is 82 to 236 years, with individuals reaching breeding size after 25 to 72 years.

Now let's introduce the slop factors: Estimated weights may be off by as much as 2 or 3 times, and if we use endothermic (warm-blooded) animals as a

FIGURE 39. Only in exceptional cases are dinosaurs known from nearly complete growth sequences. These three *Protoceratops* specimens are part of a nearly complete series of specimens collected by Museum paleontologists. Such series tell us much about the changes in form that occur during an animal's life, but they tell us nothing about how fast these animals grew or how long they lived.

FIGURE 40 (OPPOSITE) Ichthyosaurs were fishlike reptiles whose fossil remains are always found in rocks of marine origin. Although ichthyosaurs became extinct by the end of the Cretaceous and are popularly associated with dinosaurs, they were not dinosaurs and were not closely related to dinosaurs. This ichthyosaur apparently died while giving live birth. The skulls and vertebral columns of baby ichthyosaurs can be seen within the mother's body cavity and outside the body. There is no evidence that any dinosaur reproduced by live birth.

standard, instead of the ectothermic animals used by Case, growth rates are ten times higher, so we must multiply the values by 0.1. The variance in Case's rates need to be modified to reflect these values. These considerations do not take into account the real fatal flaw of Case's analysis: the eggs of *Protoceratops* have been conclusively demonstrated to belong to a different dinosaur (*see* "The Western Gobi," page 197); and there are also problems regarding the reference of eggs to *Hypselosaurus* (*see* "Eggs of ?*Hypselosaurus priscus*," page 184).

Empirical examination of the problem gives a result with so much variance that the result is uninteresting. In other words, we don't have strong enough evidence on growth rates to run an accurate scientific test, but this is the best evidence we have. Such examination is an example of both the use and the limitations of the scientific method. In this example the result is relatively uninformative, but it is nevertheless superior to the authoritarian, unsubstantiated claims often made in popular dinosaur books.

Question #32

Were dinosaurs aquatic?

In a strict sense there are no fully aquatic dinosaur species known. Some large extinct aquatic reptiles often are called dinosaurs, but they lack the features present in the common ancestor that defines dinosaurs (*see* "What are dinosaurs?," page 2) and are more closely related to other animals than they are to dinosaurs. The most familiar of these reptiles are plesiosaurs and ichthyosaurs, fully aquatic animals with modifications similar to those of aquatic mammals such as whales. Instead of limbs they had large flippers, some had heterocercal (shark-like) tails, and many had nostrils on top of their heads. A third group often

FIGURE 41. The Mayan Ranch trackway from the Cretaceous Glen Rose Formation in Texas shows what appears to be a swimming sauropod, using only the forefeet to propel itself. In contradiction to the swimming sauropod theory, however, some students of footprints have reinterpreted this trackway as underprints from an overlying trackway in which only the forefeet have pushed through the mud.

confused with dinosaurs is the mosasaurs—marine lizards closely related to the varanid lizards, whose members include the Komodo dragon.

Some nonavian dinosaurs, such as hadrosaurs, may have been partly aquatic. Many early paleontologists considered them to be predominantly so. Recent work has shown that although these animals were primarily terrestrial, some may have inhabited swampy lowlands. Some small shreds of evidence suggest that dinosaurs occasionally swam. One famous trackway of a giant sauropod preserves imprints of only front feet, initially suggesting that this animal was floating, pushing itself along with its front feet (*see* "The Paluxy River Trackway," page 181).

Dinosaurs remains in many parts of the world are found near the shores of ancient seas. Occasionally fossil sea creatures such as barnacles encrust some of these bones. This evidence does not mean that dinosaurs were aquatic, only that their bones ended up in an aquatic environment. The most aquatic dinosaurs that we have evidence for are penguins. Penguins display aquatic modifications, such as flippers and a torpedo-shaped body, and have been seen hundreds of kilometers out to sea. Unlike the ichthyosaurs, however, which fossil evidence indicates gave live birth as whales and dolphins do, penguins must return to land to lay eggs and rear their young.

Question #33

Did dinosaurs fly?

No flying nonavian dinosaurs have been discovered. The pterosaurs (another highly successful and diverse group of flying Mesozoic reptiles) are not dinosaurs, but rather close relatives of dinosaurs because they lack the features all dinosaurs inherited from their common ancestor (*see* "What are the closest relatives to dinosaurs," page 2).

With more than ten thousand species, birds are the most diverse group of dinosaurs known. How flight arose in these animals is a hot topic of debate.

Two hypotheses have been proposed. One theory, the "trees-down" hypothesis, is that bird ancestors were tree dwellers and developed powered flight by gliding from branch to ground. The converse argument, the "ground-up" hypothesis, is that birds developed powered flight directly from a cursorial (running) ancestor.

Evidence supporting the trees-down hypothesis relates to the inference that early birds appear to have lived in trees. This evidence comes from analysis of the beautiful specimens of *Archaeopteryx* that preserve not only the foot bones, but also the horny claws (equivalent to your fingernails or the claws of a cat). The claws of *Archaeopteryx* are long and hooked. In living birds this type of claw is typically found in birds that live in trees, suggesting that *Archaeopteryx* may have had similar habits. By examining the wing and shoulder, functional morphologists have inferred that *Archaeopteryx* was probably a not a highly skilled flyer. Hence, if it was also a tree dweller, true powered flight probably developed through an intermediate phase that involved gliding down from trees.

FIGURE 42. Since dinosaurs are now considered to include birds, obviously dinosaurs did and do fly. Aside from birds, however, no known dinosaurs could fly. But the Mesozoic skies were populated with other flying reptiles, the pterosaurs. Although closely related, these amazing animals are not dinosaurs.

The ground-up hypothesis is based on phylogenetic evidence. Proponents of this theory point out that the closest relatives of birds are dinosaurs like *Deinonychus* and *Troödon* (*see* "Why are birds a kind of dinosaur?," page 11). Because they are so closely related to birds, these dinosaurs provide direct evidence about physical characteristics of early birds. Although these dinosaurs were small by typical dinosaur standards, no one has proposed that they lived in trees or flew. Instead, they are viewed as fast-moving runners and denizens of open areas. These observations suggest that primitive birds may have been ground dwellers that developed powered flight from active running and hopping over open areas. Some have even proposed that the extension and flapping of the forelimbs developed as a mechanism for swatting at prey!

Unfortunately, collecting evidence to test hypotheses about the origin of complex functional behaviors such as flight is notoriously difficult. Probably the most reasonable strategy is to combine all the relevant data (from both modern birds and fossils) into a phylogenetic approach, but even this approach may not succeed, and the problem may remain intractable until new, more powerful methods of study are developed.

Question #34

What was the last dinosaur?

Because birds are still with us there is not yet any such animal as the "last dinosaur," but even determining the last of the nonavian dinosaurs is a difficult task. Much of this problem is related to the issues surrounding dinosaur extinction that we discuss in the next section. Here it will suffice to say that the Hell Creek Formation in eastern Montana and western South Dakota is one of the only rock units that documents the end of the Cretaceous period, the final chapter in nonavian dinosaur evolution.

Many types of dinosaurs have been discovered in the Hell Creek Formation, including *Tyrannosaurus*, *Pachycephalosaurus,* and *Anatotitan.*

The ceratopsian dinosaur *Triceratops*, however, is more common than any others, and *Triceratops* bones occur closest to the boundary that marks the end of the Cretaceous period. Whether *Triceratops* was the last surviving nonavian dinosaur is not certain. Because dinosaur remains are generally rare near the boundary, it is more probable that the most common types of dinosaur fossils will be found there.

Imagine a group of a hundred animals. Ninety-nine are deer and one is a cougar. Suppose that from this group, four animals are preserved as fossils. The odds are very low that the cougar would be sampled. The odds grow even lower if only one animal is preserved. Such is probably the case with *Triceratops* and *Tyrannosaurus*. *Triceratops* is the most common animal in the Hell Creek beds, so as fossils grow rarer near the boundary other kinds of animals are less likely to be found as fossils, even though they probably lived side by side with *Triceratops*.

Question #35

Why did nonavian dinosaurs become extinct?

Remember that dinosaurs are not really extinct. Birds are dinosaurs. All other dinosaurs became extinct about 65 million years ago. Why? Nobody knows for sure. Many ideas have been proposed, but scientific tests to decide which idea is the most reasonable are difficult to conduct.

For example, one hypothesis is that nonavian dinosaurs became extinct because they got hay fever from flowering plants, which arose and became common during the Cretaceous. We have no way to determine if dinosaurs got hay fever. As far as we know, living dinosaurs do not, and fossils tell us nothing about this idea.

Other ideas can be tested and rejected. One, for instance, is that dinosaurs got too big. This hypothesis is not supported by the fossil record, which shows no progressive increase in size throughout the Mesozoic era. And many relatively small dinosaurs continued to thrive to the end.

Another idea involves our own early relatives. Paleontologists who study fossil mammals are often proud to implicate their "pets" as the nemesis of the nonavian dinosaurs. They suggest that nonavian dinosaurs were eradicated either because early mammals out-competed nonavian dinosaurs for food and other resources, or because the mammals ate all their eggs. Even in modern environments competition effects are difficult for scientists to study. So understanding a worldwide ecologic system existing 65 million years ago is next to impossible because of the incomplete fossil record. What is known suggests that mammals were rare, rat-sized animals during the Cretaceous period.

Dinosaurs clearly demonstrated their ability to cope with mammals over the long run. They coexisted with nonavian dinosaurs for at least 100 million years before the extinction event. The competition scenario makes little sense, since the great diversification of mammals occurred after the demise of the nonavian dinosaurs. If anything, the fossil record shows that dinosaurs out competed mammals during the Mesozoic era.

Two contemporary ideas involve either gradual, Earth-based causes or catastrophic, extraterrestrially triggered causes. Both suggest that the extinctions of dinosaurs, along with many other plants and animals both on land and in the seas, resulted from severe changes in the climate, but the hypotheses differ concerning the duration and cause of these changes. There is evidence supporting both theories. Therefore, the question becomes, "Can we distinguish the possible effects of a catastrophic, extraterrestrial event from a gradual, Earth-based event?"

Paleontologists have long argued that the demise of the nonavian dinosaurs was caused by climatic alterations associated with slow changes in the positions of continents and seas resulting from plate tectonics. Off and on throughout the Cretaceous, large shallow seas covered extensive areas of the continents.

Data from diverse sources, including geochemical evidence preserved in seafloor sediments, indicate that the Late Cretaceous climate was milder than today's. The days were not too hot, nor the nights too cold. The summers were not too warm, nor the winters too frigid. The shallow seas on the continents probably buffered the temperature of the nearby air, keeping it relatively constant. At the end of the Cretaceous, the geological record shows that these seaways retreated from the continents back into the major ocean

FIGURE 43. Because dinosaurs are still with us as birds, we can not yet talk about a last dinosaur. Furthermore, because of the vagaries of the fossil record, determining which of the nonavian dinosaur species lived the longest, either previous to or beyond the Cretaceous-Tertiary boundary, is impossible. The bones of *Triceratops* from the Hell Creek Formation in western North America, however, represent the dinosaur whose bones have been found closest to the clay that signifies the end of the Mesozoic.

basins. No one knows why. Over a period of about 100,000 years, while the seas pulled back, climates around the world became dramatically more extreme: warmer days, cooler nights; hotter summers, colder winters. Perhaps nonavian dinosaurs could not tolerate these extreme temperature changes and became extinct.

If true, though, why did "cold-blooded" animals, such as snakes, lizards, turtles, and crocodiles survive the freezing winters and torrid summers? These animals are at the mercy of the climate to maintain a livable body temperature. It's hard to understand why they would not be affected, whereas nonavian dinosaurs were left too crippled to cope, especially if some dinosaurs were "warm-blooded" (*see* "Were dinosaurs warm-blooded?," page 52). Critics also point out that the shallow seaways had retreated from and advanced on the continents numerous times during the Mesozoic, so why did the nonavian dinosaurs survive the climatic changes associated with the earlier fluctuations but not with this one? Although initially appealing, the hypothesis of a simple climatic change related to sea levels is insufficient to explain all the data.

Volcanism, has also been implicated in dinosaur extinction. The end of the Cretaceous coincided with a great increase in volcanic activity throughout the world. Lava flooded large areas of India, and explosive eruptions in the South Atlantic and the midwestern United States hurled ash over much of the globe. These eruptions could have spewed great quantities of poisonous gases into the atmosphere, causing acid rain and more-acidic waters in the surface layers of the ocean. Over the short term, a cooling would result from the airborne volcanic ash, which would cut off sunlight. Over a longer term, warming could have resulted from the greenhouse effect. Climatic changes from increased volcanism may have caused the extinction of many nonavian dinosaurs, but they do not satisfactorily explain the selective patterns of extinction in the fossil record.

Dissatisfaction with conventional explanations for nonavian dinosaur extinctions eventually led to a key observation that, in turn, has fueled a decade of vigorous and often vitriolic debate. Many plants and animals that became extinct at the end of the Mesozoic disappear abruptly as one moves from older layers of rock documenting the end of the Cretaceous up into younger rocks representing the beginning of the Cenozoic. Between the last layer representing the end of the Cretaceous and the first layer representing the start of the Cenozoic, there is often a thin bed of clay. Scientists felt that they could

FIGURE 44. This section of rock shows the Cretaceous-Tertiary boundary as it exists in the Hell Creek Formation of eastern Montana. The dark gray sediments and the dark coal layer are from the Cretaceous. These rocks contain a typical Mesozoic fauna, rich in nonavian dinosaur remains. They are capped by a thin, light-colored clay layer that is rich in iridium. This iridium layer is thought to represent the remains of a giant comet or meteorite that crashed into Earth approximately 65 million years ago. Above the clay layer is the Tullock Formation, which preserves a typical Paleocene fauna, devoid of nonavian dinosaurs.

get an idea of how long it took to deposit this 1 centimeter of clay by looking at the concentration of the element iridium (Ir) in it.

Ir is no longer common at Earth's surface. Because it usually exists in a metallic state, it was preferentially incorporated into Earth's core as the planet cooled and consolidated. Ir is found in high concentrations in some meteorites, in which the solar system's original chemical composition is preserved.

Even today, microscopic meteorites continually bombard Earth, falling on both land and sea. By measuring how many of these meteorites fall to Earth over a given period of time, scientists can estimate how long it might have taken to deposit the observed amount of Ir in the boundary clay. These calculations suggest that a period of about 1 million years would have been required. On the basis of other evidence related to Earth's magnetic field at the time of the extinction, however, it was believed that nonavian dinosaurs and other animals had to have gone extinct within a period of half a million years. If so, the deposition of the boundary clay could not have lasted 1 million years, and the unusually high concentration of Ir seemed to require a special explanation.

Consequently, scientists hypothesized that a single large asteroid, about 10 to 15 kilometers across, collided with Earth, and the resulting fallout created the boundary clay. Their calculations show that the impact kicked up a dust cloud that cut off sunlight for several months, inhibiting photosynthesis in plants; decreased surface temperatures on continents to below freezing; caused extreme episodes of acid rain; and significantly raised long-term global temperatures through the "greenhouse effect." This disruption of the food chain and climate would have eradicated the nonavian dinosaurs and other organisms in less than 50 years.

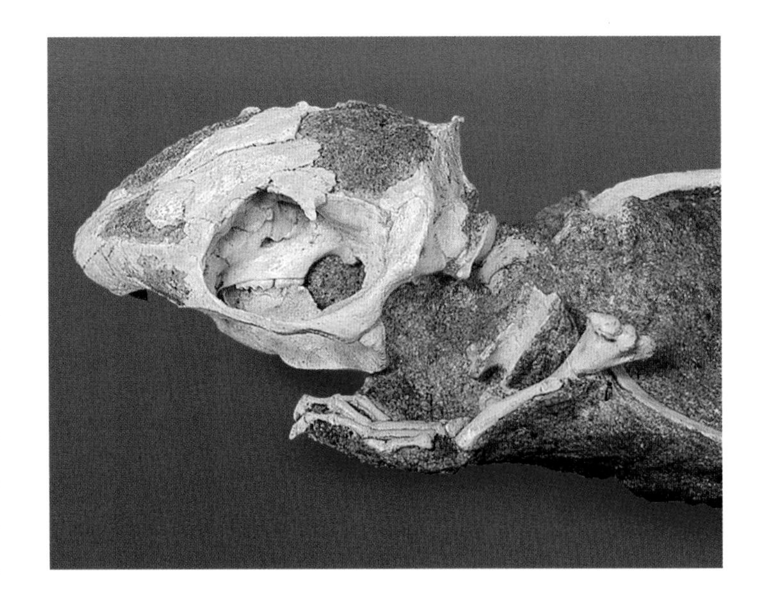

FIGURE 45. Popularly, dinosaurs are believed to have died out as a result of competition with mammals such as this multituberculate from the Gobi Desert found by American Museum of Natural History paleontologists in 1993. This theory is not credible, however, because early kinds of mammals are found nearly at the same point in the fossil record as the first dinosaurs.

To test this explanation, we must ask several questions: (1) Did an impact occur? (2) If so, which mechanism extinguished each group of organisms? (3) Can we tell time well enough to distinguish between short-term, catastrophic events and longer-term, more gradual events?

Boundary clays high in Ir have now been found in more than fifty places around the world. In rock layers that contain dinosaur fossils, the boundary clays occur about 3 meters above the highest known nonavian dinosaur fossils. Since rock layers are laid down one on top of the other, this "gap" suggests to critics of the impact idea that nonavian dinosaurs were extinct before the impact occurred. Proponents have responded that, since nonavian dinosaur fossils are uncommon, finding any fossil directly under the boundary clay would be highly improbable. Nonetheless, extensive searches have thus far proved fruitless.

For almost every line of evidence cited in support of the impact idea, critics have proposed a volcanic mechanism to explain the same evidence:

1. High Concentration of Ir. This observation seemed to represent unequivocal evidence for an impact. Subsequently, however, scientists measured the amount of Ir in gases coming out of a Hawaiian volcano. The Ir concentration in the volcanic gases is ten thousand times higher than in the lava, suggesting that large volcanic eruptions could explain the high concentration of Ir in the boundary clays. The extensive episode of volcanic activity at the end of the Mesozoic make this idea plausible.

2. Chemical Composition of the Clay. The overall chemical composition of the boundary clay (specifically the form of the element Osmium) in some localities is very similar to the composition of certain meteorites. To our knowledge, no volcanically based argument has yet been raised to counter this claim. At many localities, however, the chemical composition of the clay is closer to that of rocks formed deep within Earth than it is to that of meteorites. Critics also note that the chemical makeup of boundary clays from different areas of the globe varies, suggesting that the source was not a globally mixed dust cloud from an impact of uniform composition.

3. Minerals Fractured by the Shock of the Impact. A few beds of the boundary clay contain microscopic grains of quartz that have small fractures oriented in several different directions. Proponents argue that the fractured texture could have been created only at the temperatures and pressures generated by an impact. Less complex but similar features, however, have been recognized in some rare volcanic rocks, and some scientists believe that the Mount St. Helens eruption in 1980 was strong enough to generate the necessary temperatures and pressures to create such fractures.

4. Glassy Spheres. Impact proponents have noted the presence of extremely small spheres in the boundary clay at several sites. They argue that these represent cooled drops of molten rock that were thrown into the atmosphere by the impact. Critics counter that these may instead represent fossilized algae, volcanic ejecta, microscopic meteorites, or organic remains, especially since similar spheres are found in clay beds other than the boundary clay.

5. Soot. Soot is common in the boundary clay at Stevens Klint, Denmark. Impact advocates suggest that the soot came from massive global forest fires ignited by the fallout from the impact. Critics argue, however, that the soot may represent normal levels over a long period from forest fires set by natural causes, including volcanic activity.

6. An Impact Crater. Throughout the 1980s, scientists studied geologic data and satellite imagery searching for the "smoking gun," or, more appropriately, the "festering wound," left by the alleged impact. Earth has long been bombarded by extraterrestrial trash. Bodies large enough to pass through the atmosphere without burning up hit Earth with tremendous force leaving craters. Scars of these impacts remain clearly visible, such as Arizona's Meteor Crater. A meteor as large as the one that is thought to have caused the extinctions should have left an enormous crater, but where?

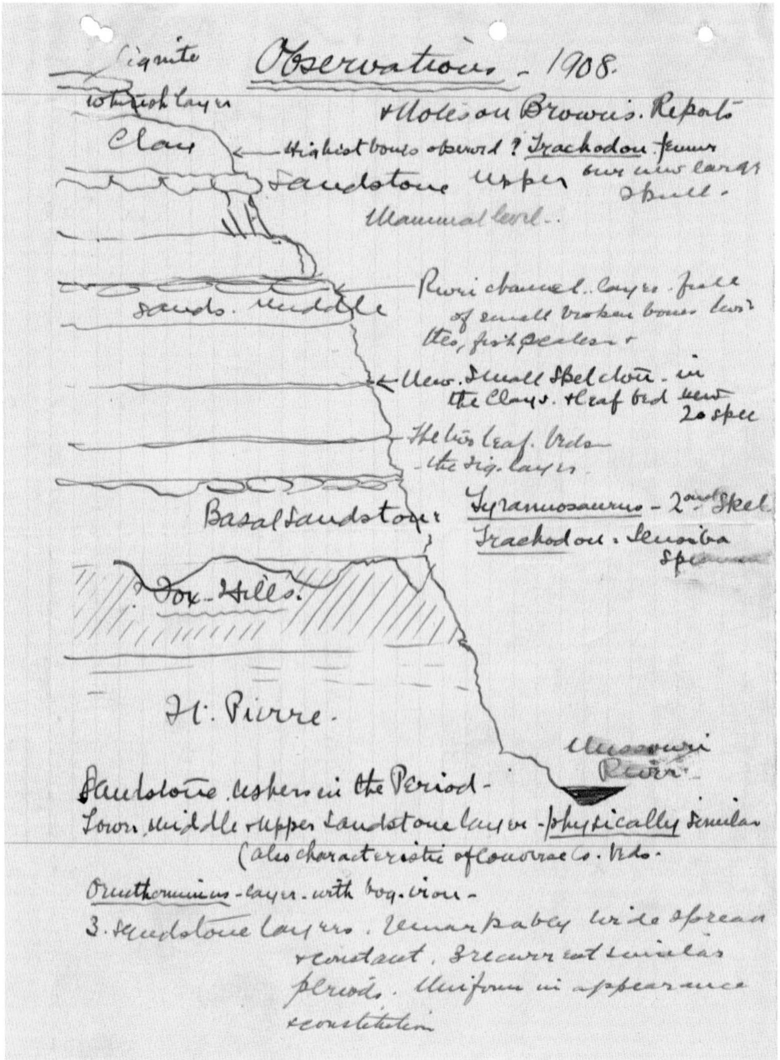

For several years, no one could find a crater of exactly the right age. Many potential explanations were proposed. Perhaps the crater was deeply buried, distorted, or destroyed by subsequent tectonic activity.

A serious candidate was finally found in the early 1990s—the Chicxulub crater on the Yucatan Peninsula and in the adjacent Gulf of Mexico. This crater is enormous, with an estimated diameter of 180 kilometers. Radiometric dating of melted rock within the crater has established its age at 64.98 million ± 50,000 years. Microscopic glassy spheres thought to be generated by the impact and found in Haiti at the same stratigraphic level as the extinction of marine microorganisms used to identify the end of the Cretaceous have been dated at 65.01 million ± 80,000 years. The most recent radiometric dating from volcanic units near the Cretaceous-Tertiary boundary just above the highest nonavian dinosaur fossils in Montana place the age of the boundary at about 65.01 million ± 30,000 years or 65.17 million ± 40,000 years. Thus, there is now good evidence to demonstrate that a extraterrestrial impact did occur at the end of the Mesozoic.

The debate between proponents of gradual, Earth-based causes and those of catastrophic, extraterrestrial causes remains heated and unresolved, although there is growing and pervasive evidence that an impact occurred. Even so we are a long way from identifying which effect triggered by the impact eradicated a particular group of animals or plants. As the number of lethal mechanisms increases and the scenario becomes more complex, testing the hypothesis and pinpointing the mechanism(s) responsible becomes that much more difficult.

FIGURE 46. These notes by Henry Fairfield Osborn, constructed in 1908 from Barnum Brown's field reports on the Hell Creek Formation, demonstrate how for nearly a century the stratigraphic relationships of dinosaur species and the Hell Creek Formation have been crucial sources of evidence for paleontologists' attempts to understand the timing and causes of the disappearance of non-avian dinosaurs.

How well we can "tell time" at the end of the Cretaceous directly affects our ability to test these competing ideas scientifically. Extinction mechanisms associated with the impact are thought to have operated over a period of less than 100 years, perhaps as little as several months. Testing this theory would require distinguishing between events that occurred less than 100 years apart more than 65 million years ago. Can we tell time that precisely?

The only direct method for determining the age of rocks and fossils is radiometric dating. These estimates are based on atomic processes of radioactive decay (*see* "How do we estimate the age of dinosaur fossils?," page 81). New technology utilizing lasers is improving the precision of dates to between 50,000 and 100,000 years for rocks formed at the end of the Cretaceous, meaning that our ticks on a clock are minimized by 50,000 years. Such error factors clearly prohibit us from distinguishing between events that occurred less than 100 years apart.

If we overlook error factors associated with radiometric dates, we can approach the question another way. Geologists have developed a method to estimate how often sediment and fossils are preserved in layers of rock. Since layers of rock are not deposited continuously, gaps of time may exist between the layers. The most complete land based sequence containing dinosaur fossils is in the San Juan Basin of New Mexico, where fossils and sediment are thought to have been preserved at least every 100,000 years. This frequency of preservation is adequate to test for extinction mechanisms lasting 100,000 years or more. For testing catastrophic, extraterrestrially triggered mechanisms that operated for 100 years or less, however, the estimates are more pessimistic. We can expect only about one out of every seventy 100-year intervals to be represented by sediment and fossils. Thus, our chances of being able to test the predictions of a single impact are slim.

Such estimates do not mean that an impact did not cause the extinctions of nonavian dinosaurs, nor that volcanic activity and the retreat of seaways did not cause the extinctions. It means that our limited ability to tell time at the end of the Cretaceous prevents us from distinguishing between an extinction that occurred quickly (less than 50 years) or over 100,000 years, at least for now. What caused the extinction of the nonavian dinosaurs? We can believe what we want to believe, and although we can collect perplexing evidence, we cannot test these ideas precisely enough to establish their scientific validity.

Question #36

What other animals became extinct at the same time as nonavian dinosaurs?

The extinction event at the end of the Cretaceous was not limited to the nonavian dinosaurs. On land, however, the Cretaceous event was not catastrophic; among vertebrates, only nonavian dinosaurs and pterosaurs were hit particularly hard, although some mammals, lizards, turtles, and crocodiles also disappear from the fossil record at this time. Probably the greatest effect was on marine organisms. Ammonites (relatives of the chambered nautilus), reef-building mollusks called rudists, many species of echinoderms, and extensive numbers of single-celled organisms called foraminifera all disappeared.

As we discussed in the preceding section, definitive determination of the timing of these events 65 million years ago is difficult. We have no way of knowing whether all these extinctions occurred nearly simultaneously or were staggered over hundreds of thousands or even millions of years. Some newer evidence suggests that the extinctions of marine animals may not coincide with those on land, and most of the marine reptiles may have disappeared about 8 million years before the end of the Cretaceous.

Question #37

Has any dinosaur DNA been found?

FIGURE 47 (OPPOSITE). Many other types of animals became extinct at nearly the same time as the dinosaurs. Many of these were sea animals. This ammonite, a relative of the present-day chambered nautilus, represents one of the diverse groups that became extinct at the end of the Cretaceous. Whether extinctions in all these animal and plant groups were synchronous is the subject of intense scrutiny.

During the last few years, probably as a result of Michael Crichton's best-seller *Jurassic Park*, considerable popular attention has focused on dinosaur DNA and the possibility of resurrecting extinct dinosaurs. Laboratories around the world have attempted to isolate DNA (the material of our genes) from dinosaur fossils. Because DNA is a fragile substance that decomposes easily and dissolves in water, it is preserved in fossils only under special conditions. One of the best places to look for ancient DNA is in specimens preserved in amber.

Organisms preserved in amber are held in an airless environment, impervious to the decomposing effects of the atmosphere. The animals are preserved almost as if they died only recently. Even coloration is often preserved. Mesozoic amber is very rare. From these rare specimens, DNA from Late Cretaceous insects has been isolated. No specimens of a nonavian dinosaur have been recovered from amber deposits. One amber sample from the Late Cretaceous of New Jersey, however, preserves a dinosaur specimen: a single feather of a bird. Scientists trying to isolate dinosaur DNA have been forced to rely on poorer materials, usually fossilized bones or teeth (*see* "How do fossils form?," page 74).

Recently there have been some preliminary reports claiming to have isolated DNA from nonavian dinosaur bones. These reports, however, are not without their problems. DNA is a component of all living organisms, and DNA extraction experiments are notoriously susceptible to contamination. The biggest difficulty is trying to determine the identity of the DNA that is isolated. In the case of nonavian dinosaur bones, the isolated DNA may be from the dinosaur, or it could be from an organism that lived in the soil near or in the fossil bone, a plant root, an aerosol contaminant, or even a scientist, via a misplaced finger.

To ensure that the isolated DNA is not a contaminant, scientists must work carefully and thoughtfully. After the DNA has been isolated and sequenced (a process that yields the genetic code specified by the building blocks of DNA), the identity of the DNA fragment needs to be determined by comparison with living animals whose DNA sequences are well studied.

These comparisons require analysis using complex mathematical protocols. Many kinds of different animals must be compared. If the mystery DNA from a fossil is the DNA of a dinosaur, the results of the comparisons should indicate that the DNA is cladistically more similar to a bird than it is to any other kind of animal or plant compared in the study, because based on the extensive body of anatomical data used to reconstruct evolutionary history (*see* "Why are birds a kind of dinosaur?," page 11) birds are the closest living relatives of the extinct dinosaurs.

FIGURE 48. The only remains of a Mesozoic dinosaur preserved in amber is this single feather. It was found by American Museum of Natural History scientists in New Jersey in rocks that are 90 million years old.

Scientists probably will confirm the presence of nonavian dinosaur DNA, but that does not mean that we will ever be able to clone extinct dinosaurs as in the movies. Even if DNA extraction experiments were 100 percent successful and we were able to recover all the individual DNA fragments of an extinct dinosaur, we would still be far from cloning a dinosaur. Probably the most important aspect of this problem is ordering.

The DNA molecule is composed of subunits called base pairs, which are two smaller subunits bonded together. Each base pair forms part of a genetic message. In our bodies every individual cell has 10^9 base pairs. It is unlikely that all of these base pairs, making up what scientists call an entire genome, could be extracted from fossil remains. Even if they could, they would still need to be assembled into an ordered, structured genome.

To draw an analogy, think of the Manhattan phone book. It has an order (alphabetical) and a content (names and addresses). If your phone book was shredded into millions of tiny pieces, each of varying lengths, how would you go about reassembling the original volume? The problem would be confounded if many of the pieces were missing and you didn't know which ones. This scenario is akin to the problem encountered by a molecular biologist who is trying to reconstruct the original genome of a dinosaur from fossil DNA.

At present, isolating and organizing the DNA into an entire genome for a fossil animal is impossible. We cannot create carbon copies of organisms that are alive today, even if we have the entire genome in its correct order. Before cloning becomes possible, much must be learned about translating the information in the genome into a living, breathing organism. As for living dinosaurs, we best be happy with pigeons, pheasants, and ostriches, because we are a long way from making *Tyrannosaurus rex* a zoo animal or seeing any other live nonavian dinosaurs.

Question #38

Did any nonavian dinosaurs survive the end of the Cretaceous?

FIGURE 49. This Andean condor is a descendent of a lineage that survived the Late Cretaceous. Dinosaurs like condors are more closely related to *Tyrannosaurus rex* than *T. rex* is to familiar dinosaurs such as *Apatosaurus, Triceratops,* and *Stegosaurus.*

Except for birds, we have no concrete evidence that any dinosaurs survived the end of the Cretaceous. The last nonavian dinosaurs appear in sediments capped by the boundary that forms the end of the Cretaceous period. Occasionally, fossil dinosaur bones appear in rocks of the Cenozoic era. Because the bones are extremely eroded and not found in articulation, however, they are considered to have been secondarily deposited, meaning that after they became fossils they were eroded and reburied in younger rocks. Such occurrences can easily trick paleontologists trying to determine the age of specific sites and specimens, but no nonavian dinosaur fossils have been unequivocally shown to be primarily deposited in rocks younger than the end of the Cretaceous.

Tabloids are quick to report alleged "Lost World" occurrences of living dinosaurs. Although these articles may make interesting reading while waiting in a supermarket line, no one has presented a shred of credible evidence that any nonavian dinosaur walks the planet today. At best, these observations represent sightings of rogue hippos, crocodiles, or decomposing marine creatures, or the simple imaginations of P. T. Barnum-style print journalists.

Question #39

What is a fossil?

Fossils are the naturally preserved physical traces of dead organisms. Most fossils represent the hard parts of the organisms, such as bones, teeth, shells, or wood. Skin imprints, trackways, and even coprolites (feces) are also considered to be fossils. Although there is no single correct definition of a fossil, objects made by prehistoric humans, such as pottery or arrowheads, are not considered to be fossils.

Question #40

How do fossils form?

Fossilization involves the replacement of the original skeleton with other material. Usually this replacement occurs when a bone or tooth is buried in the ground. Slowly the original material (calcium phosphate) is replaced by minerals (often silica) that are carried by ground water through cracks and fractures in the surrounding rock. This process of replacement can occur with complete fidelity, resulting in fossilized bones that retain their original shape and exhibit microscopic features, such as the canals that contain small blood vessels or the cavities inhabited by bone-secreting cells (*see* "Were dinosaurs warm-blooded?," page 52). Bones from the Cretaceous of southwestern Mongolia (*see* "The Flaming Cliffs," page 206, and "The Western Gobi," page 209) are some of the best-preserved dinosaur bones. Many of

FIGURE 50. These fossil birds of the genus *Presbyornis* from the Green River Formation of Wyoming date from the Eocene. Concentrations of these fossils may have resulted from catastrophic death and burial in a storm or from volcanic ashfall.

these bones still have hollow centers, like those of living dinosaurs (birds). Because rocks containing the fossils are subjected to large amounts of heat or pressure, however, bones and teeth are usually distorted during fossilization, and microstructure is obscured. Specimens can be so distorted that even paleontologists can be fooled by the identity of an element or can argue over whether a particular hunk of rock is a fossil or not!

Differences in mineral composition in rocks surrounding fossilizing bones can also modify the color of fossil bones as minerals precipitate out of the ground water. This phenomenon explains why fossil dinosaur skeletons in museums are different colors. Even specimens from the same geologic units can be different colors. The Morrison Formation (*see* "The Medicine Bow Anticline," page 198) is a Late Jurassic rock unit in western North America that has produced hundreds if not thousands of dinosaur skeletons. Most dinosaur bones from the Morrison Formation are shiny black, like those of the

Apatosaurus and *Allosaurus* specimens in the American Museum of Natural History's Hall of Saurischian Dinosaurs, but this is not always the case. Bones from the Morrison Formation at Dinosaur National Monument, such as those of *Barosaurus* (*see "Barosaurus,"* page 105), are usually light brown. Sometimes the color of fossil bones varies even on single specimens. The Museum's *Tyrannosaurus rex* is a good example. These differences in color reflect slight differences in the composition of the rock that surrounded the specimen.

Question #41

Why are dinosaur fossils so rare?

Not only dinosaur fossils are rare; fossils of all land animals are rare. The formation of a fossil requires a unique and highly improbable sequence of events. Following death, an organism must escape destructive decomposition, which usually requires rapid burial.

Think about animals today. When one dies, it is a potential candidate for fossilization. If the animal was the victim of a carnivore's attack, it will not be a very complete fossil because much of the carcass, including the bones, will be eaten, crushed, or scattered. If the animal died because of disease, old age, or a catastrophic natural accident, the skeleton, although initially intact, could be subjected to depredation by scavengers, such as vultures, hyenas, insects, and microorganisms. In addition, destructive physical forces, such as temperature changes and abrasion from moving water or wind, contribute to the low probability of fossilization.

The science of understanding fossilization in terms of decomposition and preservation is called taphonomy. Taphonomists seek to understand why some kinds of fossils occur more frequently in the fossil record than others.

FIGURE 51. After an organism dies, its chances of becoming a fossil are slim. If this recently dead bird is going to become a fossil, it must quickly be buried to protect it from scavengers and environmental influences. Even if it is entombed in rock, the rock itself must not be exposed to extremes of heat or pressure that will distort the fossilized animal, making it unrecognizable.

A taphonomist may ask why the skulls of sauropods are so rare, when so many sauropod bones are known? The answer may relate to the fragility of the skull in relation to other parts of the body and how well the skull is attached to the neck. By examining the carcasses of large animals in East Africa, taphonomists have shown that the neck ligaments of large animals are easily separated from the carcasses after death. Furthermore, because the skulls are extremely fragile (except for the teeth), they are the body parts least likely to be completely preserved.

Rapid burial of an animal after death does not ensure its preservation. The rocks entombing the fossil must also be preserved for millions of years. Geologic processes on Earth are dynamic continua of creation and destruction. Rocks can be destroyed millions of years after their formation by the erosional forces of wind and water, or by tectonic activity. Tectonism is the process in which pieces, or plates, of Earth's surface (the crust) move relative to one another, causing the crust to stress or strain. These processes are reflected on Earth's surface by mountains, earthquake faults, and deep ocean trenches. In two of the most common types of tectonic movement, pieces of the crust slide past or override each other. Such activity results in intense heat and pressure that modify rocks and destroy or obscure fossil remains.

Because of ongoing erosion and tectonic activity, older fossils generally are less common than younger ones, because older rocks have had more time to be destroyed. However, many other factors also contribute to the relative rarity of fossils of different ages (*see* "How many different kinds of dinosaurs are there?," page 15). Most important is the amount of rock deposited during a particular time interval. Because some intervals of time are represented by rock exposed at several localities, these are likely to have more dinosaur-producing rocks.

Even if the rocks preserving a fossil survive the effects of erosion and tectonics, the fossil may be impossible to find and collect. Fossils are found only when tectonic activity has brought them to Earth's surface and erosion exposes them. Many prime fossil-bearing units are covered with forests, are deposited in low-lying areas with no exposed outcrop, or are buried deep under younger rocks. During mining operations or at large construction sites, these fossils occasionally are encountered hundreds of meters underground. Although probably tens of thousands more dinosaur skeletons are preserved, at present most are inaccessible.

When paleontologists prospect for fossils, they look for small fragments of bone weathering out of the surrounding rock (*see* "How are dinosaur fossils collected?," page 83). Even a completely preserved specimen may be fragmentary if much of it has already weathered out from the ground before it is found. Rarely, maybe only once in the career of a professional fossil hunter, you will round a bend and see a small fragment of bone, perhaps the end of a tail, a single claw, or the tip of a nose. As you begin to dig you find the specimen is going into the hill, becoming more exposed as you proceed. Then the years of patient walking have paid off.

Question #42

Where is the best place to find dinosaur fossils?

Earth's rocks are divided into three general classes on the basis of how they were formed: igneous, sedimentary, and metamorphic. Igneous rocks form from molten magma that solidifies either on or below Earth's surface. Sedimentary rocks form when layers of sand, mud, or organic material settle on Earth's surface, after being eroded and transported by wind or water. Metamorphic rocks are either sedimentary or igneous rocks that have been modified by heat or pressure generated by tectonic activity. The type of rock constitutes an important clue when searching for dinosaur fossils.

Almost all fossils are preserved in sedimentary rocks. Most sedimentary rocks were deposited in ocean basins, and consequently these deposits preserve only a few dinosaur bones that were washed into the basin. Rocks that most commonly preserve dinosaur bones were deposited on floodplains by rivers or streams, at the bottom of ponds or lakes, or together with accumulations of wind-blown volcanic ash or sand.

The most important thing to know when prospecting for dinosaur fossils is where sedimentary rocks likely to contain dinosaurs are exposed on the

FIGURE 52. Dinosaur fossils are usually found in areas where there are large exposures of rocks of the right age and environment. Arid regions with Mesozoic rocks deposited in freshwater rivers, streams, and swamps yield the most dinosaur fossils. Here, Barnum Brown examines an outcrop during a collecting foray in the 1930's.

surface of Earth. In parts of the world that have been extensively studied by geologists, this is easy. A good scientific library provides access to geologic maps that show the locations of particular types of rock. In areas less well known, deciding where to look is more difficult.

Even after an area that contains sedimentary rocks is identified, finding places where these rocks are well exposed can be a challenge. Traditionally, the search has been conducted on foot, by motorized vehicle, or on horseback. Gradually, these methods are being supplanted by the study of aerial photos or high-resolution satellite images. After appropriate outcrops have been identified, the next stage is to find some fossils. The first fossils one finds will help to determine the age and the environment of the rocks. If the fossils are from marine animals, the site is probably not a good place to look for dinosaurs, because dinosaurs are not known to have lived in ancient oceans. If the fossils tell us that the rocks are only 40 million years old, the site is probably not a good place to look, because the nonavian dinosaurs died out 65 million years ago, 25 million years before these rocks were deposited.

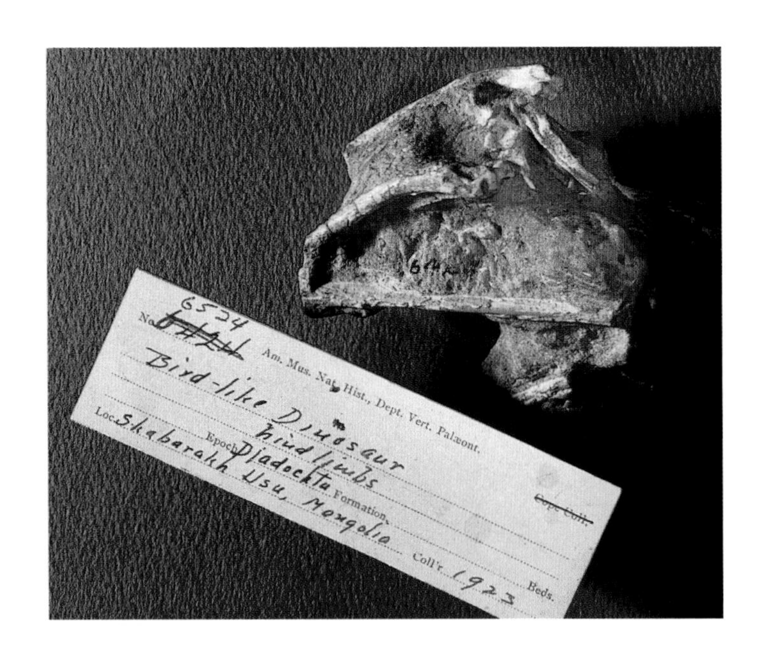

FIGURE 53. **Some of the most important dinosaurs have been found not in the field, but in museum collections. This skeleton of *Mononykus*, from the 1923 expedition to the Flaming Cliffs, was discovered in the American Museum of Natural History's collection 70 years after it was first collected.**

The best areas to find dinosaur fossils are in places where others have not looked. Consequently, collecting dinosaur fossils takes paleontologists to some of the most remote places on the planet. Very few of these remain. Mounting expeditions to these regions is arduous, expensive, and time-consuming, and there is no guarantee of finding important new dinosaur fossils. Poorly explored places are poorly explored for good reasons. And sometimes reaching a fossil-bearing region is impossible. Finally, even after months of preparation, war, disease, food shortages, weather, or political upheaval may prevent expeditions from proceeding. Many places wide open for dinosaur exploration remain, however, including Central Asia, the Sahara, and Antarctica.

Probably the least romantic place to find new dinosaurs, but one of the best, is in existing museum collections. Most expeditions collect more than can be studied within the lifetime of the excavators. Major museums have extensive holdings, some of which will be found to represent new types of dinosaurs when they are eventually prepared and studied. Even well-studied specimens often reveal new information when they are reprepared or examined using new techniques like CAT scanning. Sometimes these turn out to be new species.

In summary, the most likely places to look for nonavian dinosaur fossils are outcrops of fine-grained sedimentary rocks deposited in the terrestrial realm, between 65 and 225 million years old, and exposed in an extensive area lacking much vegetation. A temperate climate and proximity to fine restaurants with cold beer are a plus, but infrequently realized.

Question #43

How do we estimate the age of dinosaur fossils?

FIGURE 54. In the badlands of southeastern Utah different formations can be seen arranged like a layer cake, with younger formations lying on top of older ones. Dinosaurs found in these rocks can be ordered according to age depending on which formation they are preserved in.

We estimate the age of dinosaur fossils using a combination of relative and "absolute" methods. The earliest attempts involved biostratigraphy and biochronologic correlation. These approaches provide only relative ages for different fossil animals. Biostratigraphy is based on the principle of superposition, which states that younger fossils generally occur above older fossils, because layers of sediment containing the fossils are deposited one on top of the other. On the basis of this principle, the relative ages of different animals occurring together in the same rock sequence can be established. Biochronologic correlation means "telling time with fossil animals." This approach incorporates the principle of superposition and the fossil animal's stage of evolution. If similar fossil organisms are found in different rock sequences, then the fossils and rocks are of similar or equivalent age because the fossils represent about the same stage of evolutionary history.

By combining these two approaches, paleontologists collecting fossils in the nineteenth century succeeded in dividing geologic time into distinct intervals, such as Triassic, Jurassic and Cretaceous. These methods give us information about relative ages for example, that *Apatosaurus* and *Allosaurus* appear earlier than *Tyrannosaurus* and *Triceratops*. Until the middle of the twentieth century,

picks and shovels are used. If the matrix is very hard (such as cemented sandstone), heavier tools such as jackhammers, rock saws, or even dynamite, may be necessary.

The first priority is to delimit the extent of the specimen to determine how much has been preserved. Unless the specimen is found near the base camp, it is usually important to try to make the block containing the fossil as compact and light as possible because it may have to be carried a great distance. Delimiting the specimen enables a quick assessment of how much of the skeleton is preserved. These observations are recorded to help preparators when work begins back at the lab. The block is then covered with a thin layer of tissue paper or newspaper. Finally the entire block is encased in plaster of Paris bandages to form a cast, just like the casts that protect a fractured leg or arm.

After the plaster dries, wedges are driven under the cast, and the specimen is rolled over. This is an anxious moment for all fossil collectors. Two very bad things can happen: (1) If the excavation job was incomplete and the specimen extends farther into the ground than anticipated, broken bones

FIGURE 55. Although some of the techniques of fossil collecting have changed in the last century, many have remained the same. Here Barnum Brown uses plaster to create a cast around a section of *Diplodocus* vertebrae at Como Bluff in the 1890s. This was the first time that specimens were collected in this way. Previously collectors had simply chopped bones out of the ground, to be reassembled after transport to a museum or university. The technique of encasing delicate fossil bones in plaster casts is still used today.

will be exposed on the bottom of the cast. Usually, more excavation is then required and another cast must be made for the remaining skeletal parts. (2) The entire block containing the fossils may disintegrate, reducing the specimen to a heap of pulverized fragments. Fortunately, if the cast is properly made and the block sufficiently hardened, this rarely occurs. Assuming all goes well when the specimen is turned, the bottom of the cast is then plastered to enclose the dinosaur specimen completely before shipment to the preparation laboratory (*see* "How are fossils prepared?" page 87).)

Collecting large dinosaurs requires additional logistical planning and procedures. Often specimens are so big that it takes more than a single field season to procure them. Heavy equipment, such as bulldozers and large trucks are needed to excavate and transport the specimens. For the largest dinosaurs, trucks must be used to haul the massive quantities of plaster and water required in the excavation. Such an undertaking is very expensive; consequently, only a few very large dinosaurs have been completely excavated in recent years.

Question #45

How did the American Museum of Natural History acquire its dinosaur bones?

Almost all museums, including the American Museum of Natural History, have the same collecting procedures. The collection has been amassed through omnifarious means; almost every specimen has a different story. In general, museum collections are built through purchase, donation, exchange, and most important, field collecting.

The American Museum of Natural History purchased many specimens, especially during the early days of building the collection. Today the Museum does not actively pursue such purchases. Occasionally specimens or small

FIGURE 56. The Museum's most prolific dinosaur collector was Barnum Brown, who organized the excavation of the Howe Quarry in the Jurassic Morrison Formation of Wyoming. This quarry contained hundreds of disarticulated sauropod bones, as well as the partial skeleton of a juvenile *Diplodocus* (or possibly *Barosaurus*) that is now on exhibit in the Roosevelt Memorial Hall. Here, a group of tourists visit the site in 1934.

collections are donated as a gift or as a bequest of an estate. Usually these specimens lack the sort of detailed locality information that makes a specimen scientifically useful. Duplicate specimens, or more commonly casts of specimens, are often exchanged for specimens from other institutions.

Most of the new collections at the Museum are procured through field exploration. Five such expeditions are highlighted in this book. The Museum's collecting activity is not restricted to these, however; on average, more than one field party, hoping to add to the Museum's fossil collection, has gone out each collecting season for the last 125 years.

Acquiring specimens for the collection is regulated under a strict set of guidelines called the collection policy. This policy prohibits the Museum from accepting objects acquired illegally. Illegal activity is not restricted to specimens that have been stolen, but extends to specimens that have been collected without the required permits from national, local, or international authorities. By protocol, material excavated in foreign countries (*see* "The Western Gobi," page 209), must often be returned to the country of origin after study. In this case, exact replicas of the fossils are made. These are housed in the collection for study and exhibition, as if they were real specimens.

Question #46

How are fossils prepared?

Preparators are to fossils what art conservators are to paintings and sculptures. Both are highly skilled professionals that must be intensely concerned with how the specimens, or the art, will endure for study by future generations. When specimens are brought in from the field, they are encased in rock, called matrix (*see* "How are dinosaur fossils collected?," page 83). Fossil preparation involves removing this matrix surrounding the bone.

Because the matrix may be quite soft, like uncemented, unconsolidated sand, or as hard as concrete, preparation requires a variety of tools. The most familiar of these are dental tools, the same ones that the dentist uses on your teeth. These tools are used to carefully pick away matrix from near the bone. Other specialized types of small, carbide-steel needles are custom-made by preparators to fit the needs for working on a particular specimen. Until recently, preparators used small hammers and chisels extensively, especially on the preparation of large specimens. In the last few years these have been supplanted by mechanical equipment, scaled-down versions of common power tools. The most common of these are small grinding wheels driven by compressed air, air scribes (miniature jackhammers with diamond or carbide tips), and small sandblasters blowing streams of glass beads or baking soda.

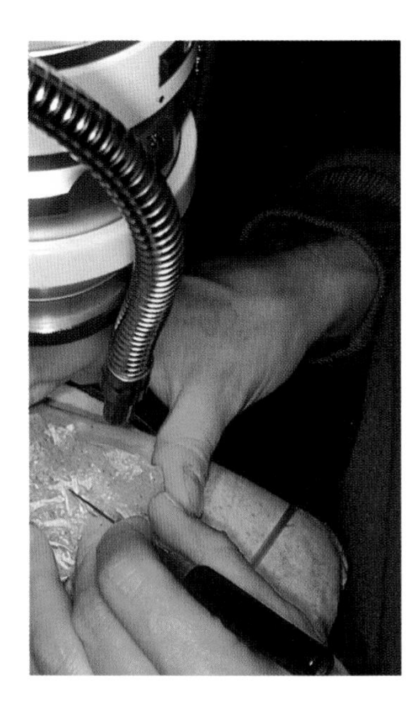

FIGURE 57. Most dinosaurs, especially small delicate ones, are prepared by hand. This painstaking process requires a good microscope, sharp tools, a steady hand, and limitless patience. Preparation of a single specimen may require hundreds of hours of difficult and demanding work.

All preparation is extremely delicate. Good lighting and high-quality microscopes are required. Often fossilized structures smaller than 50 microns (about the width of a human hair) need to be examined, so they must be separated from the matrix. Compounding the problem is the unfortunate reality that fossil materials are often much softer or more fragile than the matrix. Considering the difficulty of the work, a rare and important specimen may take months of continuous effort to prepare, even if it is very small.

The choice of materials used to strengthen or repair specimens is crucial. Even substances as inert as plaster eventually age and lose their structural integrity. Thus, preparators must think not only about how the specimen looks when it is finished, but also how it will stand up to years of study or exhibition. Several types of adhesives, glues, and fillers are commonly used to repair fractures or breaks. Generally, these substances meet the same type of standards as those used in art conservation. Brand names, compositions, and even samples are recorded and preserved to aid future preparators when the need arises to further prepare or repair the specimen.

An entirely different technique is acid preparation. Fossils are often contained in a matrix of limestone, a sedimentary rock consisting chiefly of calcium carbonate that can be dissolved by acids. A fossil preserved in limestone can be submerged in a weak bath of acetic acid to dissolve the matrix. The trick is to keep the acid from destroying the fossil. Preparators protect the fossil by coating the exposed bone with an inert polymer, such as epoxy. The specimen is placed in the acid for a short time, removed, rinsed with water, and any newly exposed bone is coated. This process is repeated until only the coated fossil material remains. The coating is removed using one of many organic solvents, and the specimen is soaked in a solution of baking soda to neutralize any residual acid. Acid preparation is tricky; many specimens have been destroyed using this technique. In the hands of a master preparator, however, even the most minute details can be freed from the limestone. Specimens prepared in acid are among the most beautiful and scientifically valuable fossils available.

An exciting new breakthrough for paleontology is CAT scan technology. Although only in its infancy, this technology promises to modify the methods of fossil preparation and study. A CAT scan is like a three-dimensional, digital X ray that produces an image of the fossil on a computer screen. It enables investigators to look through the matrix at the entombed fossil without risking damage through physical preparation. This technology allows analysis of important hidden features, such as structures inside the skull. It also enables preparators to design a preparation strategy based on the particular specimen. Although CAT scan technology probably will never totally replace traditional preparation, it is a powerful new tool for research.

Question #47

How are dinosaur mounts assembled in museums?

Most dinosaurs that are mounted today for display in dinosaur exhibitions are constructed from casts (usually fiberglass or plaster) of the original material. Mounting a cast is easy; the lightweight parts are assembled just as parts of a toy model are. Mounting real bone is much more difficult and expensive.

In the dinosaur exhibits at the American Museum of Natural History, we have tried whenever possible to utilize real fossils in our displays. This approach is a departure from exhibitions at most other natural history museums, where predominantly casts have been displayed. On display at our Museum is the greatest collection of actual dinosaur specimens ever assembled. We view this collection as akin to the collection of an art museum such as the Louvre, where

storeroom that houses many of the largest specimens on steel racks that extend from the floor to the ceiling. Many of our unprepared specimens (still in their original field jackets) (*see* "How are dinosaur fossils collected?," page 83), are stored in an adjacent room called the dinosaur annex. Finally, the smaller specimens of nonavian dinosaurs are housed in a separate room. Inside these rooms, specimens are arranged according to their taxonomic group: All the ceratopsians are kept together, and all the sauropods are kept together, regardless of where they were collected or how old they are.

As with all museum collections, individual specimens are assigned catalogue numbers. These numbers correspond to a card catalogue that records a permanent record of pertinent information relevant to the specimen. Such information includes when and where the specimen was collected, who collected it, who prepared and identified it, whether there are accompanying photographs or drawings of the specimen, and any other relevant information. These records are in the process of being computerized; eventually this information will be available electronically to researchers all over the world.

FIGURE 59. Most fossil specimens in the American Museum of Natural History are in the research collections and are not on exhibit. During a reorganization of the dinosaur collection in the 1930s, the dinosaur research collection was temporarily laid out on the floor of what is now the Hall of Saurischian Dinosaurs. Usually the collection is housed in metal cabinets and shelving, but this photograph gives an indication of the amount of research material not on exhibit.

Question #49

How accurate are artists' reconstructions of dinosaurs?

FIGURE 60. Erwin Christman produced this drawing of the skull of *Tyrannosaurus* (formerly AMNH 973). The drawing is a reconstruction of how the fossilized bones would have appeared shortly after the animal's death. This skeleton was sold to the Carnegie Museum in 1941.

What dinosaurs may have looked like in life has captured the imagination of the public since dinosaur fossils were first discovered. Some of the first dinosaurs to be reconstructed were created by English sculptor Benjamin Waterhouse Hawkins in 1853, working under the direction of Richard Owen. These were life-sized models of dinosaurs, such as *Iguanodon* and *Megalosaurus*, whose skeletal remains were commonly encountered in the English countryside. Although inaccurate, these models were so popular that the Royal Family attended a dinner to celebrate them. The success of the British exhibition led to plans for a similar exhibit in New York in 1868. The sculptures were to be exhibited in a beaux arts copy of the Crystal Palace building constructed in Central Park. In the volatile political climate of the era, however, plans for the exhibit languished, and Hawkins' sculptures were vandalized. Eventually their remains were buried under a small hill in Central Park adjacent to the Armory near the small pond at 59th Street and Fifth Avenue.

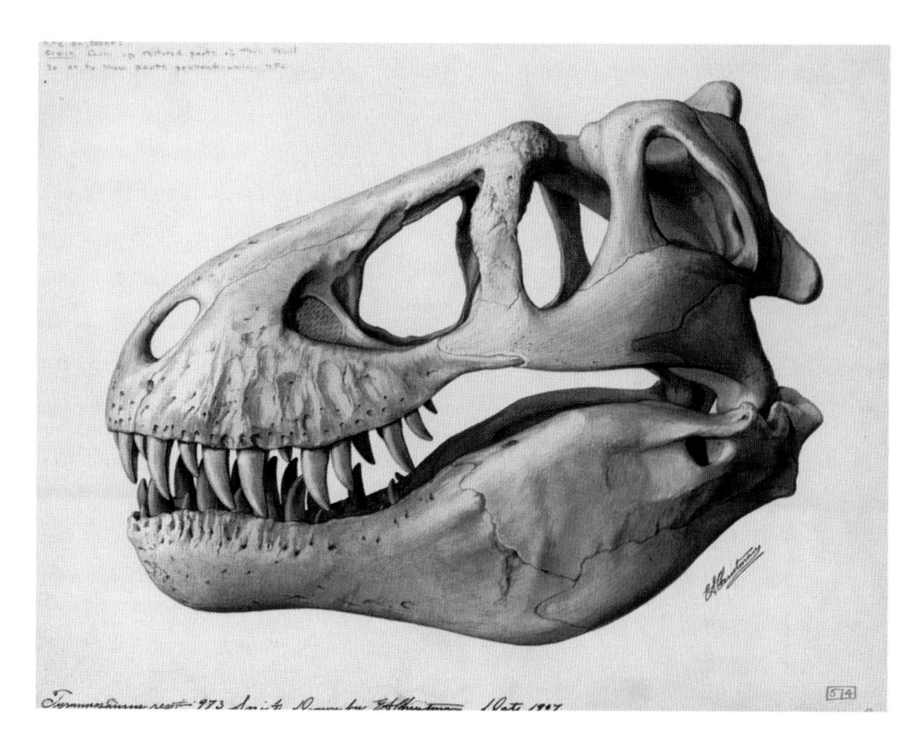

The American Museum of Natural History has a long and rich history of dinosaur reconstruction. Early in this century, Henry Fairfield Osborn contracted the services of Erwin Christman and Charles Knight as illustrators and model-makers in the Department of Vertebrate Paleontology. These men were instrumental in the early history of dinosaur reconstruction. Both were trained anatomists and were intimately involved with the work of the Museum's paleontologists. Techniques that they pioneered are still in use today.

In addition to his exquisite carbon dust-and-inkwash drawings of individual bones, Christman participated in many phases of the animal's re-creation. For many of the Museum's dinosaurs, he first reconstructed the skeleton and later the animal, as he imagined it would have appeared in life. His classic work was the lifelike reconstruction of *Camarasaurus*. To construct this model, Christman painstakingly dissected the carcasses of a wide variety of animals, carefully noting the size and orientation of muscles and bones. Next, he drew each bone of the extinct dinosaur individually, assembled them into a skeleton, added the muscles, and finally rendered the integument. Rather than painting his completed models in a lifelike fashion, Christman finished his creations with paint simulating the patina of a bronze casting. Christman's drawings and sculptures are renowned for their graceful lines and his evident mastery of technique.

Probably the most influential dinosaur reconstructor of all time was Charles R. Knight, who spent much of his career at the American Museum of Natural History. Some say that the public's perspective of dinosaurs has been influenced more by the work of Knight than by the work of any other single person, paleontologist or illustrator. Knight was one of the first to paint and sculpt dinosaurs as dynamic active animals; he often painted entire communities, replete with plants and clouded skies. Dinosaurs were not his specialty; most of Knight's best works are mammals. Nevertheless, Knight reconstructed several important dinosaurs for the Museum. Many of these illustrations reflect the way the mounts in the dinosaur halls would have looked when the animals were alive. Consequently, we have paintings of *Anatotitan* and *Allosaurus* fleshed out in the same poses as the mounted skeletons in the Museum's dinosaur halls.

Today, quality reconstruction of dinosaurs is still carried out at the Museum in the tradition of Knight and Christman. Sculptors and artists work closely with dinosaur paleontologists. Little of the technique has changed, except that materials like plaster and paper mache have been replaced by resins and plastics. As in the old days, all this work must have a firm basis in fact (the anatomy of the animals), and many modern reconstructors have become students of dinosaur paleontology.

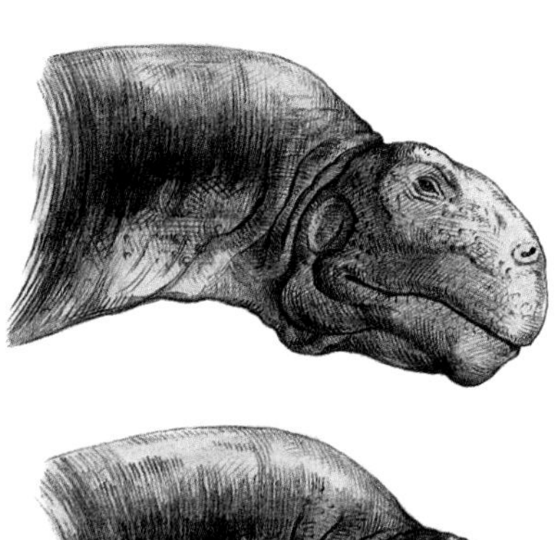

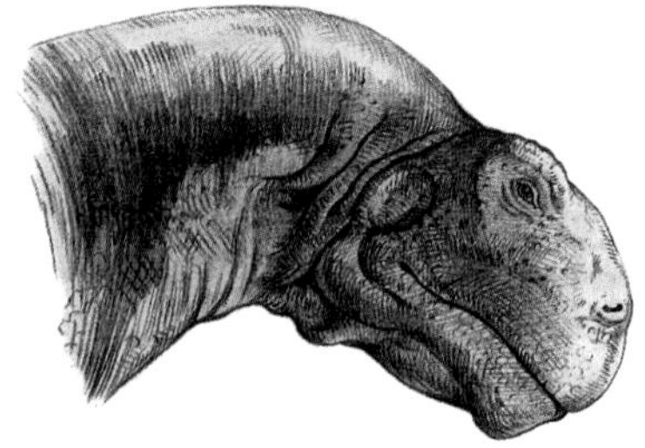

FIGURE 61. This reconstruction of the head of *Camarasaurus* was drawn by Erwin Christman for Mook and Osborn's 1921 monograph on Cope's sauropods. Christman's skill as an illustrator is unsurpassed, and his renderings are often reproduced and used as scientific tools today. Reconstructions such as this one, no matter how lifelike, are fanciful, based on pure speculation rather than on scientific fact.

Question #50

How does dinosaur study benefit modern humans?

As in justifying any other intellectual pursuit (for instance, painting, writing, or classical scholarship), dinosaur science cannot be evaluated by the monetary worth of its products. Nor can it be rated as relatively more, or less, important than medical research, computer science, or nuclear physics. Instead, it should be evaluated for what it brings to society. Without going overboard about the merits and relevance of paleontology, we would argue that it has made society richer—at least by virtue of our ever-increasing understanding of the issues examined in this book. Bringing this understanding of dinosaurs and the past into the perspective of our everyday life, we gain a heightened awareness and appreciation for our planet and its past.

One tangible outcome of dinosaur science is its role in science education. One of the buzzwords in this field is "science literacy." According to almost all contemporary sources, the public's understanding of general scientific principles is increasingly deficient. Americans, inhabitants of the economically most powerful nation in the world, are some of the most scientifically ignorant of any developed country. Yet more and more jobs require a basic understanding of high technology. An often cited problem with science education is that students consider science to be boring. As scientists, we view this more as a difficulty with presentation, rather than the relative appeal of the material. One interesting proposal has been to use the almost universal interest in dinosaurs (*see* "Why are dinosaur fossils so interesting?," page 5) to develop a curriculum for teaching basic scientific concepts. Because dinosaurs are such a diverse group of organisms, questions embracing all sorts of scientific concepts can be developed around dinosaur-related topics: for example, limits on dinosaur size and speed (physics and chemistry), the kinds of rocks dinosaurs are found in (geology), how dinosaur remains are dated (physics and chemistry), whether an asteroid caused dinosaurs to become extinct (astronomy), where dinosaur remains are found (geography), and how birds are related to dinosaurs (anatomy). This approach is being used in several universities, and dinosaur science courses are some of the most popular science classes for nonmajors. This approach has even begun to filter down into preuniversity education.

The importance of paleontology extends farther. This endeavor is the direct outcome of the intuitive curious nature of humans, as we try to understand our place on this planet. Fossils and specifically dinosaurs, have been instrumental in the development of evolutionary theory, as well as many other historical concepts, such as plate tectonics and biogeography, now taken for granted. Without these fossils, we would have no indicator of our own species' relative insignificance even during the time that humans have occupied the planet. If this does not strike one as a revelation concerning our place in nature, then it is best look to mysticism for answers.

As the end of the millennium approaches, one of our major dilemmas is how to develop concern, appreciation, and a heightened understanding of Earth's biodiversity. From this perspective, it is important to realize that today we live with only a small fraction of the total number of kinds of animals that have existed on the planet. Those remaining are a living legacy of what we collect as fossils, because all organisms are interrelated through our evolutionary heritage documented by sequences of characteristics inherited from lineages of common ancestors. Crucial to our understanding of how to preserve the diversity that remains is an understanding of these interrelationships, because the pattern of interrelationships establishes the magnitude of Earth's biodiversity. Fossils are an important part of the calculus for understanding these relationships. The way fossils and living organisms are phylogenetically interwoven underscores the fact that extinct species are not isolated entities from Earth's past. They, like we, are all integrated actors in the drama we call the evolution of life.

11 THE GREAT HALL OF DINOSAURS

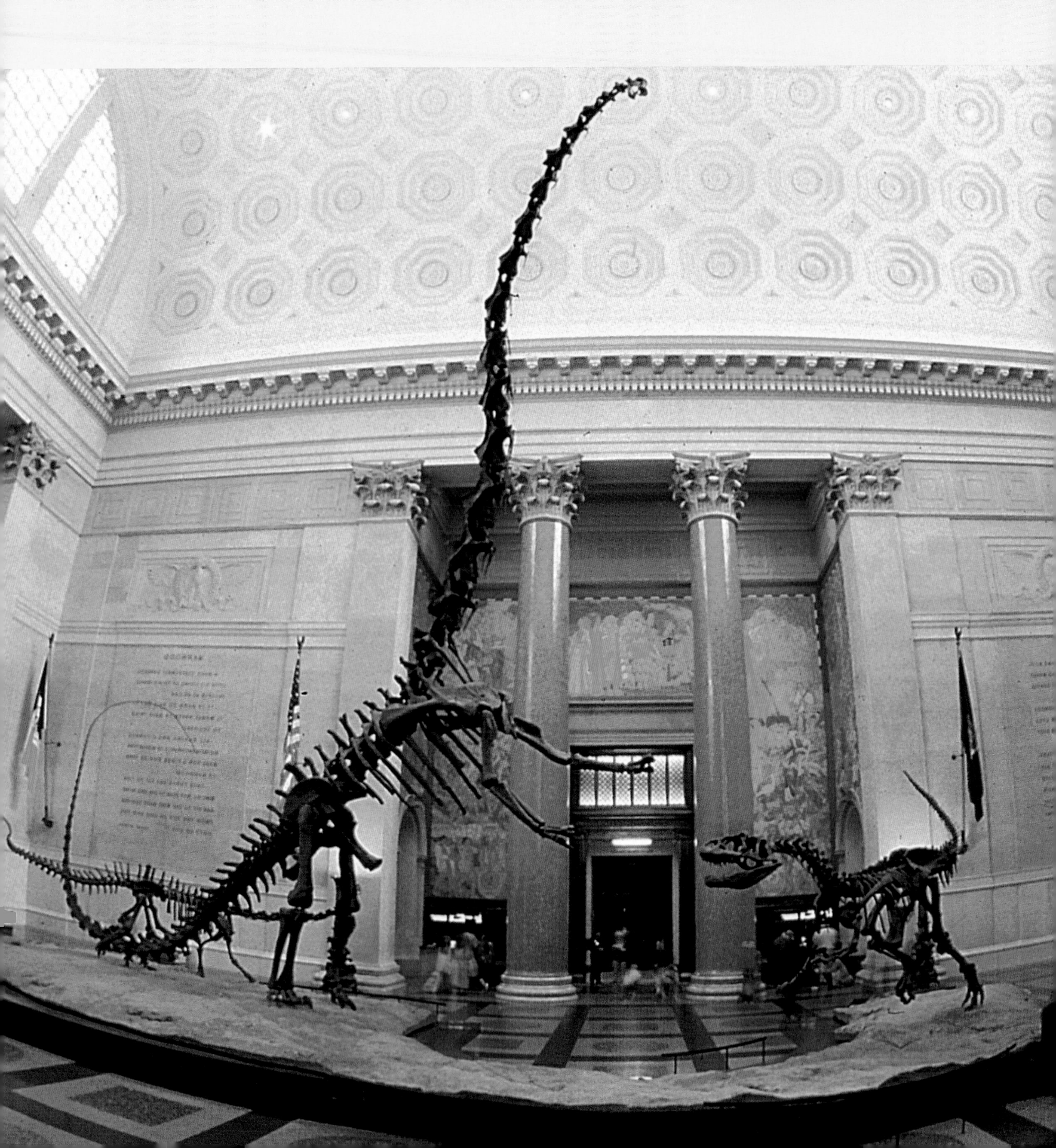

SAURISCHIAN

Plateosaurus engelhardti

ANMH: *6810*
AGE: *Late Triassic, about 220 mya*
FORMATION: *Knollenmergel*
LOCALITY: *Trossingen, Baden-Wårttemberg, Germany*
COLLECTOR: *Friedrich von Huene*
DATE: *1923*

FIGURE 63. The bones of the prosauropod *Plateosaurus* are often found in mass accumulations. This slab at the Museum was collected in a quarry in Bavaria, Germany.

Plateosaurus is the best-known member of the prosauropods, which were early relatives of the giant sauropods, such as *Apatosaurus* and *Diplodocus*. The prosauropods and sauropods are related; both possess long necks and small heads, shared derived characters that were inherited from their common ancestor. Nonetheless, prosauropods retained many features that were primitive for dinosaurs, and they are important because they show what the early dinosaurs were like. The forelimbs were smaller than the hind limbs, suggesting bipedality, the primitive locomotion pattern for dinosaurs. Trackways, however, show most prosauropods to be quadrupedal. Similarly, prosauropod teeth were arranged in single rows, in contrast to the more complex dental batteries of more advanced dinosuars such as duckbills.

Like other prosauropods, *Plateosaurus* was probably herbivorous, using its narrow, serrated teeth to slice food. Rounded pebbles, or gastroliths, in the gut formed a gastric mill to process plant material (*see "Psittacosaurus,"* page 165).

SAURISCHIAN

Camarasaurus lentus

AMNH: *467*
AGE: *Late Jurassic, about 140 mya*
FORMATION: *Morrison*
LOCATION: *Bone Cabin Quarry, Wyoming*
COLLECTOR: *W. D. Matthew*
DATE: *1904*

FIGURE 64. Despite the popular fame of *"Brontosaurus"* (a.k.a. *Apatosaurus*), *Camararaurus* was the first of the giant dinosaurs to be known from specimens complete enough to make dependable reconstructions. This drawing of *Camarasaurus* (AMNH 467) appeared in Mook and Osborn's 1921 monograph on Cope's sauropods.

Camarasaurus is one of the best-known sauropods because its bones have been found throughout Jurassic rocks of the Morrison Formation. This formation consists of sands and muds that were deposited by rivers on wide flood plains and deltas similar to the Mississippi Delta of today. *Camarasaurus* bones are often preserved in ancient sand bars and river deposits where they accumulated like driftwood. Although deposits like those at Dinosaur National Monument in Utah contain thousands of *Camarasaurus* bones, most of these specimens are only partially articulated because much of the carcass had rotted and broken apart before burial. Only a few nearly complete *Camarasaurus* skeletons are known. The best one is a 17-foot-long juvenile from Dinosaur National Monument, now in the Carnegie Museum.

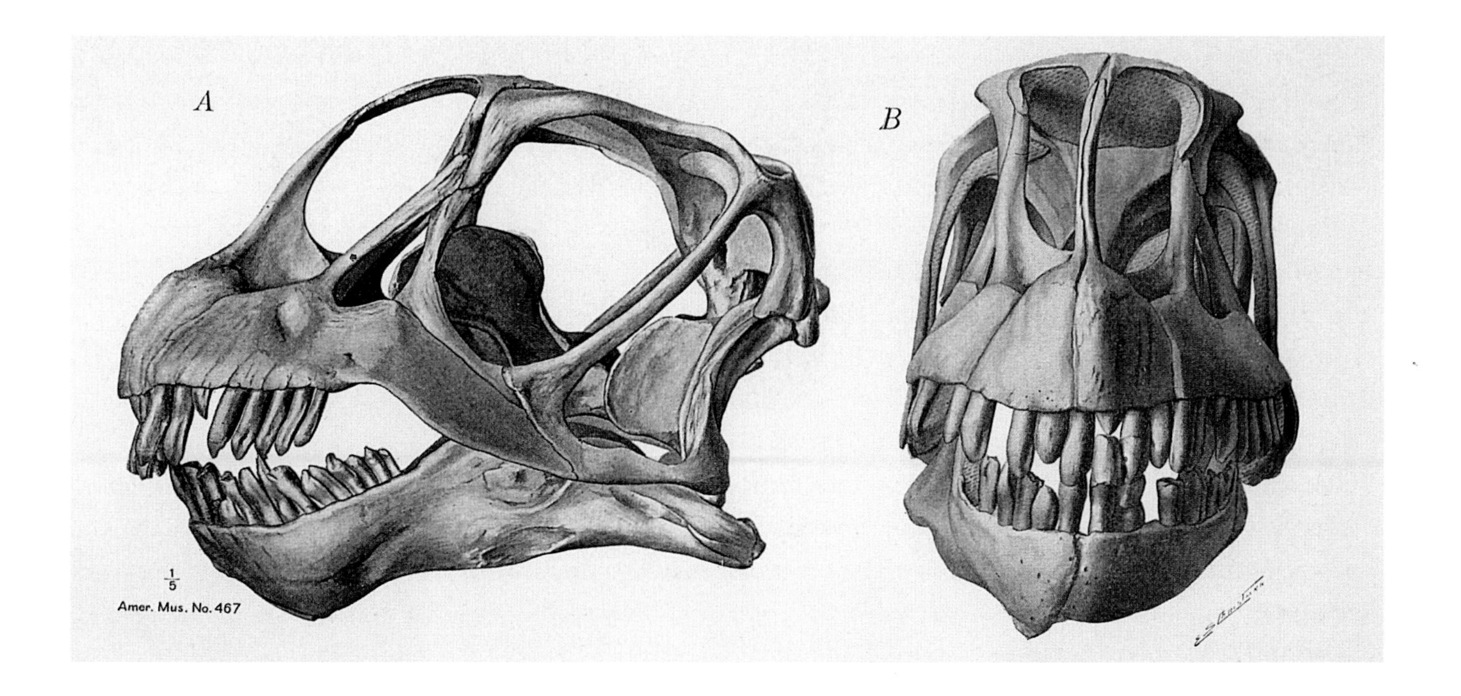

Although the *Camarasaurus* specimens housed in the American Museum of Natural History are not the most complete specimens that have been collected, they have great historical importance: These specimens formed the basis for our early understanding of sauropods. Much of the Museum's *Camarasaurus* collection was brought together by Edward Drinker Cope as the result of a chapter in the Cope-Marsh rivalry that dominated paleontology in the late nineteenth century (*see* "The Medicine Bow Anticline," page 198).

SAURISCHIAN

Diplodocus longus

AMNH: *969*
AGE: *Late Jurassic, about 140 mya*
FORMATION: *Morrison*
LOCATION: *Bone Cabin Quarry*
COLLECTOR: *Peter Kaisen*
DATE: *1903*

The *Diplodocus* specimens in the American Museum of Natural History are historically significant because they were acquired during the "Golden Age" of dinosaur collecting. In fact, the first dinosaur collected by the Museum was a partial skeleton of *Diplodocus* (AMNH 223) from Como Bluff, Wyoming, in 1897 (*see* "Who discovered the first dinosaur bones?," page 6).

Subsequent expeditions to the Como Bluff area in the early 1900s yielded some high-quality specimens from nearby Bone Cabin Quarry. This locality was famous for yielding skulls and skeletons of small dinosaurs, specimens rarely preserved in the Morrison Formation. *Diplodocus* skull material from Bone Cabin Quarry includes a well-preserved braincase. The brain of *Diplodocus* was very small by mammalian standards, but studies comparing endocranial cavities of dinosaurs with those of recent crocodilians, lizards, and turtles show that dinosaurs did not have unusually small brains (*see* "How intelligent were dinosaurs?," page 37).

FIGURE 65. The skull of *Diplodocus* is typical of diplodocid sauropods like *Apatosaurus* and *Barosaurus* in that it has small pencil-shaped teeth restricted to the ends of the jaw and nares (nose openings) on top of the skull between the eyes. Although the bones of giant sauropods such as *Diplodocus* are common worldwide, their skulls are very poorly represented in the fossil record. This lack of skull material is probably due to the relative fragility of the skulls, as well as the weak connection between the skull and the neck. This particularly fine *Diplodocus* skull (AMNH 969) was collected at Bone Cabin Quarry in 1903.

FIGURE 66. Barnum Brown's second wife, Lillian, was an enthusiastic traveler and related her life with Barnum in such works as "I Married a Dinosaur." She visited Howe Quarry in Wyoming during one of Brown's last large-scale digging operations.

Another Wyoming site, Howe Quarry, yielded more than 32 tons of bone, mostly of the sauropods *Camarasaurus, Diplodocus*, and *Apatosaurus*. The sauropod collection from Howe Quarry includes a rare juvenile specimen (AMNH 7530) that probably belongs to *Diplodocus*, but may be the related and similar *Barosaurus*. This specimen consists of the neck and skull and was used to help reconstruct the composite juvenile *Barosaurus* in the Roosevelt Memorial Hall (*see* page 97). The Howe Quarry excavation of 1934 was the last time that large sauropods were collected by the Museum.

SAURISCHIAN

Apatosaurus excelsius

AMNH: *222, 339, 460, 592*
AGE: *Late Jurassic, about 140 mya*
FORMATION: *Morrison*
LOCALITY: *Nine Mile Quarry and Bone Cabin Quarry, Wyoming*
COLLECTOR: *Walter Granger and others*
DATE: *1897, 1898*

FIGURE 67. The Museum's skeleton of *Apatosaurus,* unveiled in 1905, was the first sauropod to be mounted for exhibition anywhere. No techniques or guidelines had been developed by previous experience, so the Museum's workers who were mounting the specimen developed their own. The *Apatosaurus* forelimbs were articulated by trial and error, but subsequent evidence has shown no serious errors in their position.

Apatosaurus, formerly *called Brontosaurus* (*see* "How do dinosaurs get their names?," page 8), is one of the most familiar dinosaurs. *Apatosaurus* and its close relatives include the largest known terrestrial animals ever—some as long as 50 meters. Because there are no living analogues, the dietary habits, biophysics, and lifestyles of these animals have been debated extensively. As we emphasized in earlier discussions, very little data from the fossil record bears directly on these issues.

The American Museum of Natural History's *Apatosaurus* skeleton was the first of its kind ever mounted. The mount is a composite skeleton, consisting of bones from more than one individual, but the main part (about two-thirds) of the skeleton represents one individual (AMNH 460) from Nine Mile Quarry, Wyoming. This skeleton was discovered in 1898 by Walter Granger and excavated by the Museum's field party during the summer of 1899. The chest, neck, vertebrae and ribs were found articulated, but the other elements—pelvis, left femur, and other limb bones—were scattered. Added to this skeleton were the right femur, tibia, and scapula of AMNH 222, tail vertebrae of AMNH 339, and the foot bones of AMNH 592. The remainder of the skeleton is modeled in plaster from an *Apatosaurus* in the Peabody Museum at Yale University.

The preparation and mounting of *Apatosaurus* took a team of preparators, working full time, until 1905 to complete. The mounting involved two controversies: what the skull of *Apatosaurus* looked liked and how sauropods stood. To this day no skull has been found directly attached to an *Apatosaurus* skeleton. Osborn thought *Apatosaurus* was related to the camarasaurs and had a plaster camarasaur-like skull put on the skeleton. Recent examination of associations between skeletons and skulls, along with the recognition that *Apatosaurus* is more closely related to *Diplodocus* than *Camarasaurus*, has resulted in a change of heads. The leading sauropod specialist, J. MacIntosh, has shown that a skull found close to, but not attached to, an *Apatosaurus* skeleton, is probably an *Apatosaurus* skull, as W. J. Holland had first claimed in 1915. In the current remount of the American Museum of Natural History's *Apatosaurus*, this skull is used.

The second controversy involved the stance. Some paleontologists thought sauropods walked like crocodiles with their legs sprawled out sideways. Others thought the legs were elephant-like pillars. Osborn and his associates dissected crocodiles and other recent reptiles and added paper muscles to the *Apatosaurus* skeleton as it was prepared. Otto Falkenbach (a Museum technician) carved a detailed wooden model of an *Apatosaurus* skeleton to permit study of different poses. Osborn rejected the sprawling stance in favor of an upright posture because the limb joints would have dislocated if the animal sprawled. Time has supported this decision; a few articulated sauropod skeletons and the discovery of sauropod trackways show that the upright stance is probably correct.

As part of the current renovation of the Museum's dinosaur halls, a section of the *Apatosaurus* mount was modified to include new information gathered in the nearly 90 years since it was originally constructed. New discoveries show that part of the neck and the end of the tail were restored incorrectly. The remounting took three people nearly a year of almost continuous work. Because this mount is of such historic value, the modification protocol required materials and techniques that give the revised mount the same look as the original. The remounting included the addition of four neck vertebrae (for a total of fourteen), replacement of the skull, suspension of the neck by thin steel cables from the ceiling, and extension of the long, dragging tail by about 7 meters. The tail was also suspended from the ceiling in a whiplike fashion high off the ground because trackways show no evidence that the tail dragged. The skeleton is now located at the entrance to the Saurischian Dinosaur Hall, where visitors are greeted by a lengthwise view of this giant animal ambling toward them.

FIGURE 68. The revitalized *Apatosaurus* mount shows the tail suspended high off the ground. Evidence from trackways whick lack a furrow for the tail, support this interpretation.

SAURISCHIAN

Barosaurus lentus

AMNH: *6341*
AGE: *Late Jurassic, about 140 mya*
FORMATION: *Morrison*
LOCALITY: *Dinosaur National Monument, Utah*
COLLECTOR: *Earl Douglass*
DATE: *1912-1914*

What do we really know about dinosaurs? This is the major theme of the American Museum of Natural History's dinosaur exhibit renovation of the 1990s. To express this theme in an exhibition, we have developed a mount stretching the limits of objective knowledge, prompting the visitor to ask, "How do you know this is true?" The Museum chose to portray a scene from the Jurassic, about 140 million years ago. An adult *Barosaurus* rears up on its hind legs to protect a juvenile *Barosaurus* from attack by a hungry *Allosaurus*. Could this have taken place? Sure. Is there objective evidence that scenes like this actually did take place? No.

Although the closest living relatives of sauropods—birds and crocodiles—show varying degrees of parental care, there is very little evidence that sauropods stomped on predators in defense of their young. Some sauropod trackways, however, show evidence of different size classes traveling together, implying a social structure (*see* "Did dinosaurs travel in herds?," page 45). The idea of an adult *Barosaurus* rearing up on its hind limbs is not new. Before the turn of the twentieth century Charles Knight (*see* "How accurate are artists' reconstructions of dinosaurs?," page 93), working under the direction of Henry Fairfield Osborn, depicted a large sauropod rearing up to feed. Nevertheless, we have no footprint evidence suggesting that sauropods reared up on their two hind limbs. Data from close relatives are not applicable, because the relevant aspects of crocodile and bird physiques are not comparable to those of giant sauropods.

Many arguments in the scientific community, both pro and con, have focused on the ability of sauropods to rear up. Usually these questions are cast in a biomechanical perspective. Could the hind limbs have supported the immense weight of the trunk, forelimbs, and neck? How could blood be pumped from the heart all the way up to the brain? A particularly foolish response to the second question was that *Barosaurus* had nine "extra" hearts located in its neck. This idea is highly unlikely because it would require an inconceivable exception to the rule followed by all other vertebrate animals of having only one heart.

Biomechanical arguments that predict behaviors and capabilities of animals from their bones are notoriously hard to test. These methods are difficult to apply even to animals alive today. Living animals often exhibit unexpected capabilities. For instance, given only one skeleton to examine, could you envision the circus antics of modern Asiatic elephants—rearing upright on their hind limbs, standing on their head, or walking on their front feet?

Structures developed in soft tissues cannot be inferred accurately from bones. Consequently, we have no record in extinct dinosaurs of structures like the heat-sink ears of elephants or the romantically crucial fantail of a male peacock. Giraffes, for example, have specialized muscles that act as pumps to aid the movement of blood from the torso up the long neck to the brain. Rather than having nine hearts, isn't it more likely that long-necked sauropods (such as *Barosaurus* and its relatives) also had a similar specialization? The only direct evidence we can muster are comparisons of structures that can be observed directly in fossil skeletons, evidence from close relatives, and trackway information.

Unfortunately, none of these sources provide good empirical evidence about whether *Barosaurus* could rear up. For our purposes (as reflected in the design of the *Barosaurus* mount), the available lines of evidence do not rule out such a stance, and it remains a spectacular, thought-provoking possibility. Perhaps the most powerful support for the idea that *Barosaurus* could rear up on two legs comes from other aspects of dinosaur biology (*see* "How did dinosaurs mate?," page 47). Even if protection of the young or hunger for leaves high off the ground could not motivate the *Barosaurus* to rear up on only two limbs, perhaps romance could.

The American Museum of Natural History's *Barosaurus* skeleton was discovered by Earl Douglass during the Carnegie Museum's extensive field operations at what is now Dinosaur National Monument in Utah. Because sauropods are such large animals, the excavators did not realize at first that all the bones they were collecting belonged to one skeleton. By the time it became apparent that only one individual of the rare *Barosaurus* was represented, three institutions—the University of Utah, the Smithsonian Institution, and the Carnegie Museum—owned the different parts. In 1929 Barnum Brown

FIGURE 62 (SEE PAGE 97). Unveiled in 1991, the grouping of three Morrison dinosaurs, and adult *Barosaurus* protecting its young from an advancing *Allosaurus*, was the first installment of the Museum's comprehensive dinosaur renovation.

made a series of trades and purchases that resulted in the Museum's acquisition of the entire skeleton. The skeleton (which is about 80 percent complete) is the most complete sauropod specimen in the American Museum of Natural History, and the most complete of the five partial *Barosaurus* skeletons known.

SAURISCHIAN

Coelophysis bauri

AMNH: *7223, 7224*
AGE: *Late Triassic, about 220 mya*
FORMATION: *Kayenta*
LOCALITY: *Ghost Ranch, New Mexico*
COLLECTOR: *George Whittaker and Carl Sorenson*
DATE: *1948*

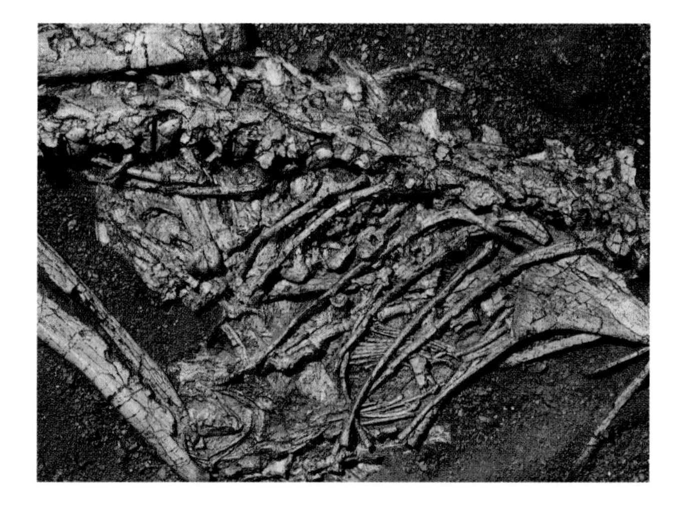

FIGURE 69. The bones of a juvenile *Coelophysis* in the stomach of an adult *Coelophysis* (AMNH 7223) tell the tale of cannibal feeding.

A major discovery of dinosaurs by the Museum involves one of the oldest and most primitive dinosaurs. In June 1947 curator Edwin Colbert and his assistant George Whitaker prospected outcrops of the Late Triassic Chinle Formation at a place called Ghost Ranch in the Chama River valley near the northern New Mexico town of Abiquiu. On the third day of prospecting, Whitaker found fossil fragments scattered on a large talus slope. Colbert immediately recognized the fossils as remains of the small, poorly known theropod *Coelophysis*. *Coelophysis* was already known from fragments found in the area, but in seventy years since David Baldwin, then a collector for Cope found the first remains at Arroyo Seco, New Mexico, no new elements had come to light. Over the next few days Colbert and Whitaker identified the location of the bone bed and realized that a vast excavation job lay ahead. What had been planned as a two-week prospecting trip metamorphosed into two complete field seasons of excavation.

In contrast to dinosaur collecting during the "Golden Age," 1895-1920, excavation of the Ghost Ranch *Coelophysis* skeletons proceeded with the aid of air hammers, compressors, and motorized vehicles. All the back-breaking, time-consuming work of reaching the bone layer, finding the limits of the skeletons, and dividing them into blocks, however, had to be done by hand as usual. During the summers of 1947 and 1948 more than a dozen intertwined skeletons of *Coelophysis bauri* were collected from a thin bed of red sandstone. In the 1980s a group of museums led by the Carnegie Institution reopened the *Coelophysis* quarry at Ghost Ranch and collected more specimens.

The Ghost Ranch quarry is one of the great dinosaur death assemblages ever collected. Here is one of the few places where we have a large sample of a single population, with both adult and juvenile individuals. Almost all the animals found here are *Coelophysis* specimens. One question is how did all of these animals die? Various ideas have been proposed, but no single theory is completely satisfactory. If we look to the present, occasionally we see large accumulations of animal carcasses massed in one place as the result of catastrophic events. For instance, a few years ago tens of thousands of caribou drowned while trying to cross a flood-swollen river in northern Canada. Along this river the bloated, decomposing bodies were stacked 3 meters high. Similar occurrences happen each year at river crossings during the annual migration of east African antelopes. Among dinosaurs such occurrences have been hypothesized at places other than Ghost Ranch, most notably in southern Alberta, where there is a massive accumulation of ceratopsian bones (*see "Centrosaurus,"* page 168).

FIGURE 70. The long, lightly built skull of the Triassic theropod *Coelophysis* (AMNH 7227) is filled with sharp teeth, corroborating the interpretation that it was an agile carnivore. Found in fine grained sandstones at a site yielding more than twenty skeletons, *Coelophysis* is one of the best-known Triassic theropods.

SAURISCHIAN

Allosaurus fragilis

AMNH: 5753, 666
AGE: *Late Jurassic, about 140 mya*
FORMATION: *Morrison*
LOCALITY: *Como Bluff region, Bone Cabin Quarry, Wyoming*
COLLECTOR: *H. F. Hubbell (for E. D. Cope); Peter Kaisen*
DATE: *1877, 1901*

A*llosaurus* is a large theropod that reached a length of 12 meters. In terms of fossil material, it is the best known of the carnosaurs. *Allosaurus* was probably an active bipedal carnivore that used the large claws on its feet and hands, as well as its sharp, serrated teeth, to overpower, kill, and eat animals as large as or even larger than itself. It is enticing to speculate on the habits of such a spectacular predator, but we cannot objectively test such ideas.

Some evidence suggests that large carnosaurs such as *Allosaurus* could swallow huge pieces of meat. The skull of *Allosaurus* was not an immobile block, but was composed of separate modules that moved in relation to one another. The skulls of many living animals, including those that eat large prey, such as monitor lizards and snakes, are also capable of such movement. Movement between the skull bones may be beneficial both for swallowing large pieces of meat and for counteracting the tensional forces caused by struggling prey. These ideas are still speculative, however, and the precise functions of various elements in the skull and limbs of carnosaurs cannot be determined with certainty.

A mysterious fossil bed containing mostly *Allosaurus* bones was discovered at the Cleveland Lloyd Quarry in central Utah. More than ten thousand bones, representing several individuals from different size and age classes, have been recovered from this site. The bones of the large predators *Ceratosaurus, Stokesosaurus,* and *Marshosaurus*, although less abundant than *Allosaurus*, have also been collected there. The large number of predators at this site has led to rampant speculation concerning how they came to be preserved in the deposit.

What is remarkable about Cleveland Lloyd is the huge number of predator remains relative to the sparse remains of plant-eating dinosaurs. At most fossil localities, the remains of carnivorous animals are much rarer than the remains of plant eaters,

FIGURE 71. This mount of *Allosaurus* feeding on a partial skeleton was one of the first dinosaur mounts constructed of an animal in a lifelike, active pose.

SAURISCHIAN

Tyrannosaurus rex

AMNH: *5027*
AGE: *Late Cretaceous, about 65 mya*
FORMATION: *Hell Creek*
LOCALITY: *Hell Creek region, Montana*
COLLECTOR: *Barnum Brown*
DATE: *1908*

Tyrannosaurus rex was the largest carnivore ever to walk on Earth and is probably the most famous of all dinosaurs. Although these animals have been the subject of much speculation, very little is known about them. Contributing to the mystique of *Tyrannosaurus* is its rarity. Only about a dozen specimens have been collected, and most are fragmentary. For more than 50 years after the original discovery, only two skeletons were known. Barnum Brown collected both of these during the early 1900s.

The type specimen (AMNH 973) consists of part of a skull, the hind limbs, and much of the vertebral column. This was the first specimen collected during the inaugural Hell Creek expedition of 1902 (see "Hell Creek Beds," page 201). On August 12, 1902, Brown wrote Osborn: *"Quarry No. 1 contains the femur, pubes, part of the humerus, three vertebrate, and two indeterminate bones of a large carnivorous dinosaur, not described by Marsh. I have never seen anything like it from the Cretaceous."* Neither had anyone else, and in 1905 Osborn formally described *Tyrannosaurus rex* based on this specimen.

Three years later Brown discovered a second, better specimen. News of this was received by an elated Osborn who wrote to Brown, *"Your letter of July 15th makes me feel like a prophet and the son of a prophet, as I felt instinctively that you would surely find a* Tyrannosaurus *this season… I congratulate you with all my heart on this splendid discovery… I am keeping very quiet about this discovery because I do not want to see a rush into the country where you are working."*

Unearthing these specimens was a tremendous undertaking. Archival photos and visits to the site show that extensive overburden needed to be removed, without the aid of power equipment. Collecting staples like plaster were in short supply, and Brown often substituted flour paste. Many of the blocks weighed over 1,500 kilograms, and these had to be transported by wagon to the nearest railhead, more than 200 kilometers of rough road away. Nevertheless, after the 1908 season, the Museum had in its possession all the

FIGURE 74 (INSET). This skeleton of *Tyrannosaurus rex*, mounted in 1915, was the first dinosaur displayed in the old Hall of Man. Later it was moved, first to the old dinosaur hall (now the Early Mammal Hall), then to the Cretaceous Hall (now the Hall of Ornithischian Dinosaurs), and finally, in 1992, to the Hall of Saurischian Dinosaurs.

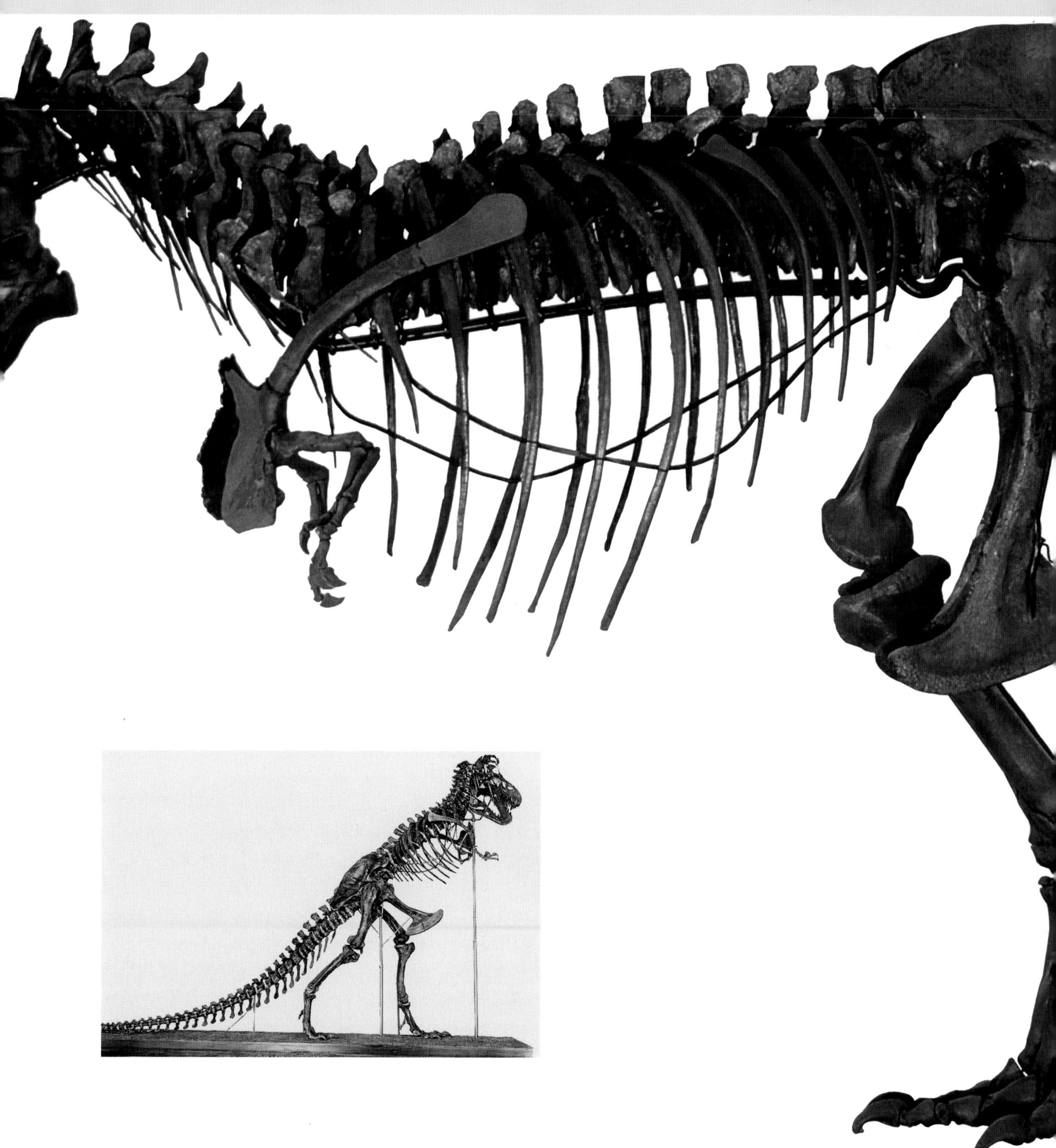

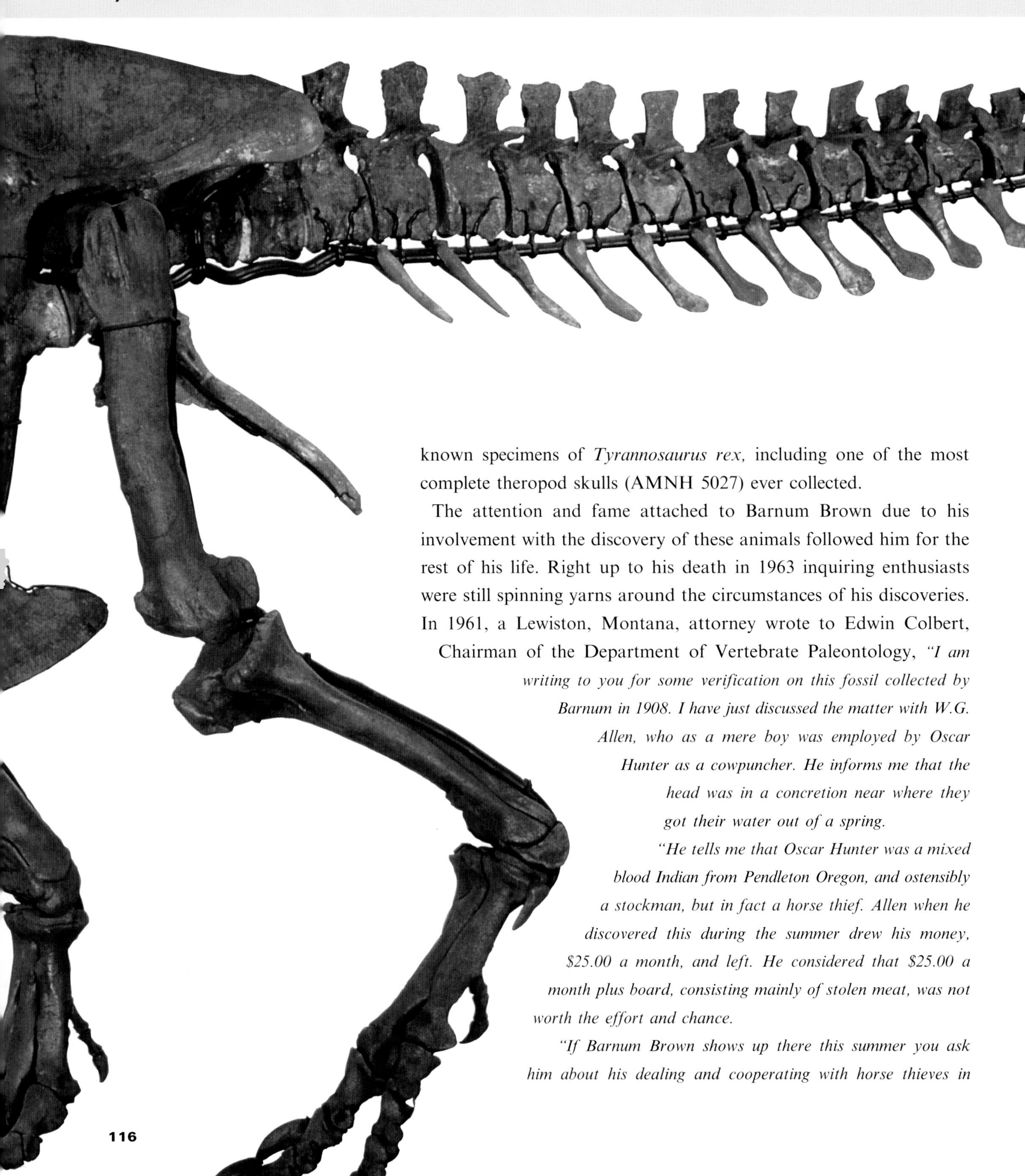

known specimens of *Tyrannosaurus rex,* including one of the most complete theropod skulls (AMNH 5027) ever collected.

The attention and fame attached to Barnum Brown due to his involvement with the discovery of these animals followed him for the rest of his life. Right up to his death in 1963 inquiring enthusiasts were still spinning yarns around the circumstances of his discoveries. In 1961, a Lewiston, Montana, attorney wrote to Edwin Colbert, Chairman of the Department of Vertebrate Paleontology, *"I am writing to you for some verification on this fossil collected by Barnum in 1908. I have just discussed the matter with W.G. Allen, who as a mere boy was employed by Oscar Hunter as a cowpuncher. He informs me that the head was in a concretion near where they got their water out of a spring.*

"He tells me that Oscar Hunter was a mixed blood Indian from Pendleton Oregon, and ostensibly a stockman, but in fact a horse thief. Allen when he discovered this during the summer drew his money, $25.00 a month, and left. He considered that $25.00 a month plus board, consisting mainly of stolen meat, was not worth the effort and chance.

"If Barnum Brown shows up there this summer you ask him about his dealing and cooperating with horse thieves in

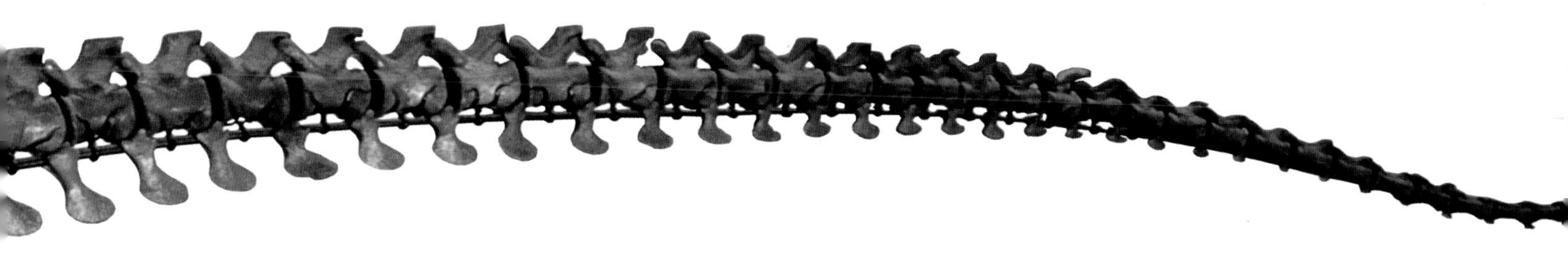

FIGURE 75. The new mount of *Tyrannosaurus rex* portrays the animal in a sleek stalking position. This pose is based on new information that indicates that like other theropods, tyrannosaurs walked with their backs horizontal to the ground and carried their rigid tails high off the ground. The tail acted as a counter-balance for the large body.

order to collect a valuable fossil. You can tell him that the statute of limitation has run out against the crime and that we are not holding it against him. It is a compliment to his ability to get along with the people and get their cooperation."

The letter was passed on to Brown who attempted to set the record straight: *"Before your letter to Dr. Colbert I had never heard that Hunter was a horse thief; that you know, if proved, in the early days was a hanging offense. The specimen you speak of as lying near a spring, I think was a trachodon skull; certainly it was not one of the tyrannosaurs."*

World War II precipitated the sale of the type specimen (AMNH 973) to the Carnegie Museum. Although this story has been hard to authenticate, Barnum Brown wrote in a note that, *"The type specimen which had most of the limb bones preserved was sold to the Carnegie Museum after we had made casts of the limbs in 1941; as we were afraid the Germans might bomb the American Museum in New York as a war measure, and we had hoped that at least one specimen would be preserved."*

In 1993 the Museum initiated plans to remount the *Tyrannosaurus* specimen in preparation for its relocation to what is now the Hall of Saurischian Dinosaurs. Because of the fragility of the bones and of the materials used in preparing the original mount in 1915 (*see* "How are fossils prepared?," page 87), a complete repreparation of the original bones was required. Many of the bones in the original mount had never been completely prepared out from the matrix. The pose proposed for the remount necessitated additional preparation. The entire procedure, only the first stage in rebuilding *T. rex*, took several people nearly a year to complete.

While conservation of the specimen was under way, a team of designers and mount-makers began to draw up plans for the supporting metal armature (*see* "How are dinosaur mounts assembled in museums?," page 89). A stalking pose was chosen. The head is low to the ground, with one foot just beginning to be lifted and the tail held high and outstretched. Engineers and metallurgists were consulted about the types of materials and the points of stress that would need to be taken into account in building the new mount.

After all the planning and conservation of the bones was complete, construction of the mount began: first the base, then the armature to support the legs and hips. An elaborate carriage and gantry were designed to support the pelvis, which weighs nearly 200 kilograms. To eliminate the heavy vertical supports that were present in the original mount, steel girders were built into the ceiling to allow most of the weight of the skeleton to be supported by steel cables. The specimen is for the most part suspended from above rather than propped from below.

Fabrication of the armature was an arduous process. A separate metal bracket had to be fabricated for each element. These brackets are similar to jewelry mounts holding precious stones: Each tine must be crafted around every contour of the fossil and is designed to enhance the dynamics of the pose. Finally, a lightweight cast of the skull (not the original, which has never been mounted on the specimen, because of its weight) was added. This cast differs noticeably from the original in that it has been completely prepared. All the fenestrae, the eye socket, and the nose were opened up, giving the skull an airy, light appearance.

FIGURE 76. This *Tyrannosaurus* skull (AMNH 5027) is the finest such specimen ever described. Although not the largest *Tyrannosaurus rex* known, this skull measures more than 1.5 meters in length.

SAURISCHIAN

Albertosaurus libratus

AMNH: *5336, 5458, 5664*
AGE: *Late Cretaceous, about 72 mya*
FORMATION: *Judith River*
LOCALITY: *The Red Deer River region, Alberta, Canada*
COLLECTOR: *Barnum Brown and Peter Kaisen; Charles Sternberg*
DATE: *1913, 1914, 1917*

FIGURE 77. *Albertosaurus* and *Tyrannosaurus*, the advanced carnosaurs, have very similar skulls with large openings in front of the eye (the second hole from the back of the skull). The lower jaws were very deep and contained jaw muscles. The teeth were replaced in waves, causing the irregular appearance of the rows, with older teeth constantly being replaced by new ones.

Fueled by spectacular new discoveries of tyrannosaurs in western North America, much attention has been focused recently on the posture and habits of these animals. Although the notion of active carnosaurs with raised tails is not new, this view has not always been prevalent, despite the supporting evidence from specimens. One *Albertosaurus* specimen (AMNH 5664) was collected in its death pose. (Death poses of many dinosaurs, especially theropods, typically show the animal with its head arched back like a dead seagull; *see "Struthiomimus,"* page 121.) When AMNH 5664 was collected, its tail was directed straight out, curving slightly upward, from the back of the animal. This pose was retained in the original Museum display specimen. When remounted for exhibition in the 1950s, however, the tail was modified to its current position, dragging behind the animal. Close examination of the specimen in its present position reveals that the contacts between tail vertebrae do not fit together well; consequently, this tail position would have been impossible in life. Furthermore, no fossilized trackways of theropods show evidence of drag marks from the tail.

Most paleontologists have likened carnosaurs to monstrous vicious predators, giving them the metaphorical name "land sharks." Others suggest that *Albertosaurus* and its relatives were large scavengers. The evidence is inconclusive, except for the ferocity, which is implied by injuries preserved in the fossil bones (*see* "Did dinosaurs fight?," page 42). From a purely intuitive perspective, it seems unlikely that these large, highly mobile animals, equipped with serrated, saberlike teeth and large claws, were simply scavengers, although they may have taken advantage of a newly dead carcass if they came across it.

Carnivorous dinosaur skeletons are very rare, but the Museum specimen of *Albertosaurus libratus* (AMNH 5664) is the most complete advanced carnosaur skeleton known from North America. It lacks only the end of the tail, most of the left forelimb, and some ribs. It was collected in 1917 by Charles Sternberg and purchased by the Museum in 1918 for $2,000. This specimen was mounted for display by Peter Kaisen and Carl Sorenson and unveiled in 1921. In 1923 Brown and Matthew described this fossil as the type specimen of *Gorgosaurus* (an invalid synonym for *Albertosaurus*) *sternbergi*, noting that this specimen was much more gracile than a typical *Albertosaurus libratus*. Since then this specimen has been considered to be a juvenile *A. libratus*. The differences in build between *A. sternbergi* and *A. libratus*, however, are striking; consequently, more study will be required before we can decide whether this specimen represents a different species or is a juvenile.

FIGURE 78. Charles Hazelius Sternberg led his family of three sons in a fossil collecting business. Later the sons associated themselves with various institutions. Here, *Albertosaurus sternbergi* (AMNH 5664), the type and only known specimen of this species, lies at the feet of Charles Sternberg.

A second spectacular *Albertosaurus* specimen (AMNH 5458) was collected by Barnum Brown, Peter Kaisen, and Albert Johnson near Steveville, Alberta. Like AMNH 5664, this specimen was mounted in a running posture. Mounts like these—of big, active animals—show definitively (in contradiction of some revisionist historians) that paleontologists of the early twentieth century thought of these animals as active, mobile, fast predators.

Several other beautiful *Albertosaurus* specimens were also collected by Museum field parties on the Red Deer River. A few of these were sent to other institutions. Notably, a well-preserved skeleton collected in 1913 and prepared as a plaque mount was transferred to the Smithsonian as part of a deal to acquire the *Barosaurus* skeleton (*see* page 105) in 1933. Today this *Albertosaurus* specimen is on display in the Smithsonian's dinosaur halls.

SAURISCHIAN

Struthiomimus altus

AMNH: *5339, 5421*
AGE: *Late Cretaceous, about 70 mya*
FORMATION: *Horseshoe Canyon*
LOCALITY: *Sand Creek, the Red Deer River region, Alberta, Canada*
COLLECTOR: *Barnum Brown and Peter Kaisen*
DATE: *1913, 1914*

Ornithomimids are called ostrich dinosaurs because their toothless beaks and gracile physiques resemble those of their distant modern relatives. Primitive ornithomimids had teeth, so the beak in birds and ornithomimids is not evidence for direct evolutionary relationship. Rather, like the wings of birds, bats, and pterosaurs, the beaks are the product of convergent evolution. The lack of teeth has led to speculation that ornithomimids ate plants, but such speculation is weak (*see* "What did dinosaurs eat?," page 39) because there are beaked dinosaurs alive today (such as vultures, hawks, and roadrunners) that eat meat.

SAURISCHIAN

Oviraptor philoceratops

AMNH: *6517*
AGE: *Late Cretaceous, about 72 mya*
FORMATION: *Djadoctha*
LOCATION: *The Flaming Cliffs, Mongolia*
COLLECTOR: *George Olsen*
DATE: *1923*

Among the weirdest of all theropod dinosaurs are the oviraptorids, of which *Oviraptor* was the first to be recognized. These are relatively small animals; the largest are the size of an adult ostrich. In general, the skeletons are quite ordinary for a maniraptor. They all have long hind limbs with sharp claws on their feet. Similarly, th front limbs are long and equipped with nimble, clawed fingers. The wrists have the curved "semilunate" carpal present in the common ancestor of all maniraptors.

What makes oviraptorids so bizarre is their heads. The upper and lower jaw margins lack teeth; a beak is present instead. The skull is very lightly built and is full of large cavities and holes (called fenestrae). The most unusual aspect of the *Oviraptor* skull, however, is the large crest present on some of the specimens. These crests may have been covered with horny sheaths in life, just as in the modern hornbill, which the skull of *Oviraptor* superficially resembles. Not all oviraptorids, even among those thought to belong to the same species, have crests. This observation has led to the suggestion that, like the horns of a moose or elk, the crests were sexually dimorphic and may have been used in display. Because of the extremely birdlike skulls (no teeth, large eyes, and crest) some paleontologists have restored *Oviraptor* with plumage similar to that of large ground birds. The discovery of *Caudipteryx* (see "*Caudipteryx*," page 188) lends credence to this view. The evidence for sexual dimorphism remains ambiguous at best. Although the problem requires more study, other maniraptors like *Velociraptor*, which are more closely related to birds than is *Oviraptor*, are predicted to be similarly feathered.

The unusual skulls of oviraptorids have sparked an extensive debate concerning the dietary habits of these animals. Almost every imaginable possibility has been proposed, including eggs (hence the name, which means "egg hunter"), small animals, plants, and mollusks. Determining the dietary preferences of animals that have been dead for tens of millions of years is extremely difficult (*see*

FIGURE 81. When found over a nest of presumed *Protoceratops* eggs, this lightly built, toothless predator was named *Oviraptor*, literally "egg robber." Recent discoveries (*see* "The Western Gobi," page 209), however, show that the eggs contained embryonic oviraptors and were therefore laid by them rather than eaten by them. Although *Oviraptor* is closely related to the birdlike theropods and birds, the more precise relations of this small dinosaur with the bizarre skull is unknown.

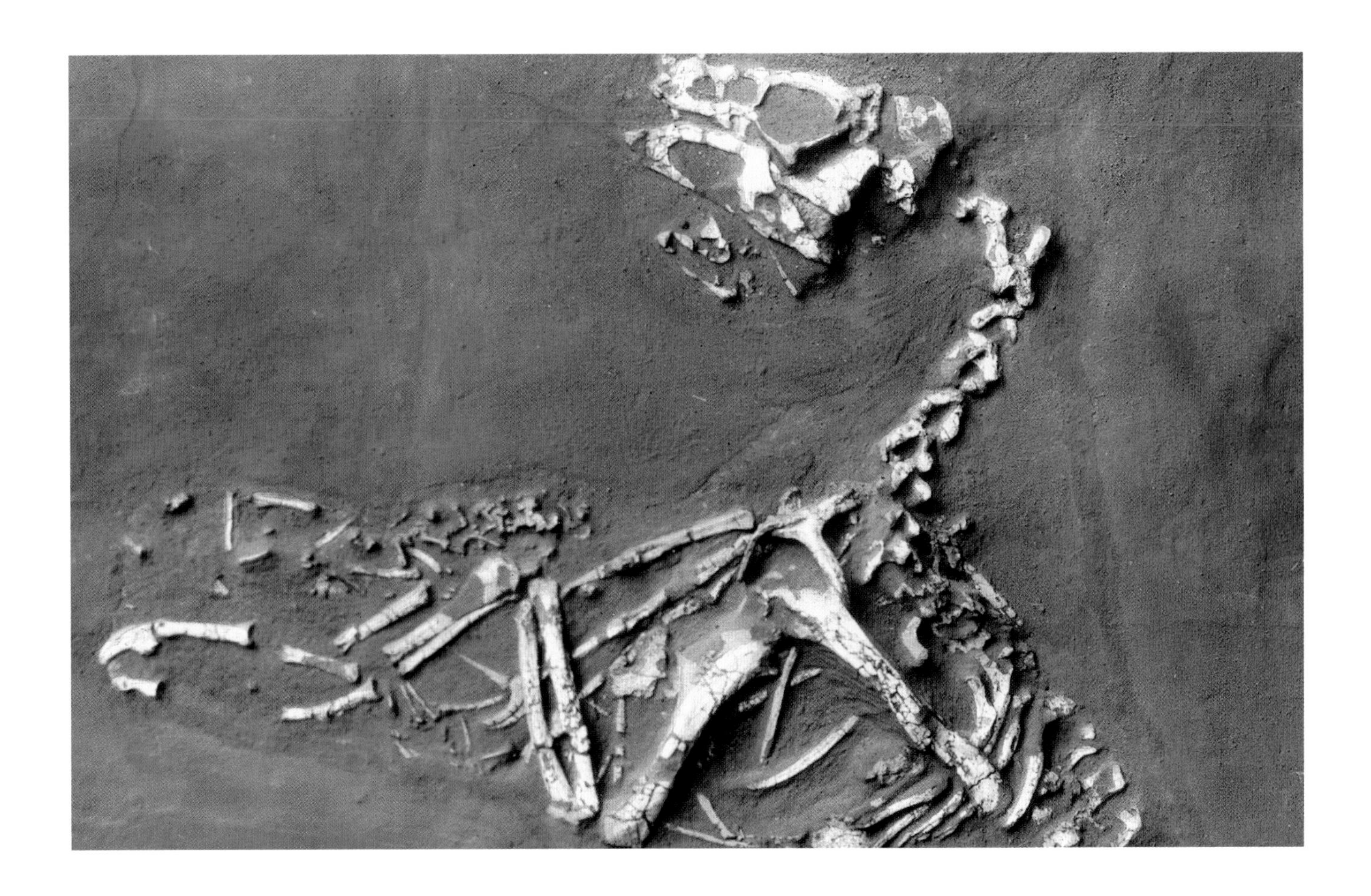

"What did dinosaurs eat?," page 39), but in this case we may have some direct evidence. Associated with the Museum's type specimen (AMNH 6517) are the bones of a small lizard near the stomach cavity. Lizards make up the diets of a number of birds, so it's not unlikely that lizards were a common prey item of these close avian relatives. This lizard may have been final meal of the *Oviraptor* specimen.

The type specimen of *Oviraptor* is a nearly complete skeleton that was found atop a nest of dinosaur eggs by George Olsen in 1923. There are two possible explanations: the animal was feeding on the eggs when it died, or it was incubating or protecting the nest. Osborn, who described the specimen, preferred the former, stating the position of the fossil *"immediately put the animal under suspicion of having been overtaken by a sandstorm in the very act of robbing the dinosaur egg nest."* Recent evidence based on new material collected by the joint expedition of the Mongolian Academy of Sciences and the American Museum of Natural History favors the latter (*see* "The Western Gobi," page 209).

SAURISCHIAN

Microvenator celer

AMNH: *3041*
AGE: *Early Cretaceous, about 110 mya*
FORMATION: *Cloverly*
LOCALITY: *Wheatland County, Wyoming*
COLLECTOR: *Barnum Brown*
DATE: *1933*

M *icrovenator* is a small dinosaur known only from a fragmentary skeleton found in the Early Cretaceous Cloverly Formation of southeastern Montana. The only specimen ever collected of this animal is the American Museum of Natural History type specimen (AMNH 3041). This specimen is very incomplete, consisting of only parts of the skull, several vertebrae, and a few limb bones. Although the evolutionary relationships of *Microvenator* are poorly understood, it has been suggested that it is closely related to, or even a member of, the Oviraptoridae.

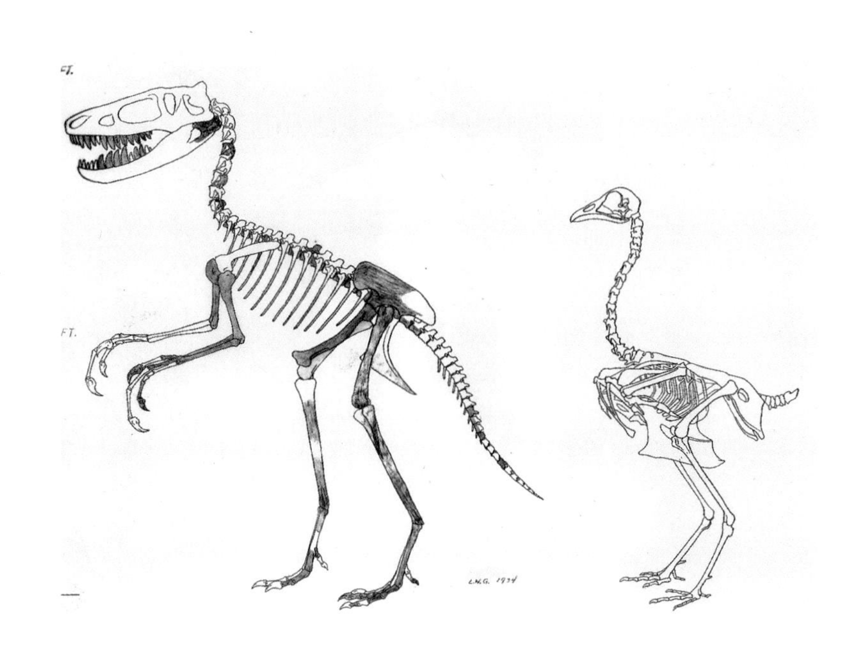

FIGURE 82. ***"Megadontosaurus"*** **is shown here compared to the skeleton af a bird, in one of Brown's unpublished figures. Fossils collected later have shown that Brown included teeth of the larger theropod** ***Deinonychus*** **with a smaller theropod,** ***Microvenator.***

At the collection site, the type specimen of *Microvenator* was found adjacent to some large serrated teeth which Barnum Brown assumed belonged to the animal. Because these teeth seemed disproportionately large compared to the small, gracile skeleton, he called this animal *Megadontosaurus* ("large teeth"). Although he had an extensive set of illustrations made for this specimen, Brown never formally published an article on the specimen, so *Megadontosaurus* is not a valid name (*see* "How do dinosaurs get their names?," page 8).

In the 1960s John Ostrom began working in the Cloverly Formation at many of the same localities that the Museum had worked during the first half of the twentieth century. Ostrom's field work produced the Yale University specimens of *Deinonychus, Tenontosaurus,* and *Sauropelta.* While examining Barnum Brown's small theropod labeled *Megadontosaurus*, Ostrom realized that the large teeth catalogued with this specimen were identical to those of

Deinonychus, a larger Cloverly maniraptor. This observation suggested that Brown's interpretation of the specimen needed revision. Ostrom proposed the name *Microvenator celer* for the smaller skeletal parts and referred the large associated teeth to *Deinonychus*.

Is it possible that Barnum Brown was right—that there was a small theropod with a huge head in the Cloverly deposits? Probably not. Finding the remains of several different animals in close proximity is a relatively common occurrence in vertebrate paleontology. The only definitive proof that all the bones at a single site belong to a single animal is if the bones are found in life-like articulation.

SAURISCHIAN

Saurornithoides mongoliensis

AMNH: *6516 (type)*
AGE: *Late Cretaceous, about 72 mya*
FORMATION: *Djadoctha*
LOCALITY: *Flaming Cliffs, Mongolia*
COLLECTOR: *Chih*
DATE: *1923*

One of the more interesting, but understudied, Mongolian theropods is the small troödontid *Saurornithoides mongoliensis*. Similar in size and anatomy to dromaeosaurs such as *Velociraptor* and *Deinonychus, Saurornithoides* and its troödontid relatives also had a second digit modified into a sharp cutting claw. This claw was not as large in troödontids, however, as in dromaeosaurs. It is now well established that both troödontids and dromaeosaurs are very closely related to birds. Even Roy Chapman Andrews, while not commenting directly on relationships, noticed similarities with birds: *"These dinosaurs are much too late in geological time to be ancestral to birds, but they do parallel them remarkably in their almost wing-like hands and lightly built skulls."*

Saurornithoides is a very rare animal. In four years of collecting, recent American Museum of Natural History expeditions (*see* "The Western Gobi,"

page 209) have failed to collect a single good specimen. Besides the original specimen (AMNH 6516), only a couple other specimens of *Saurornithoides* are known. Troödonitds from North America (*Troödon*) and Inner Mongolia (*Sinornithoides*), however, are known form larger samples of material.

Proportionally, troödontid have the largest brains of any nonavian dinosaurs. Brain size and other morphologic features have been cited as evidence that they were highly intelligent animals with acute sensory capabilities. For instance, troödontids had extremely large eyes, which are displaced laterally, perhaps allowing fields of view to overlap. These overlapping fields of view allow stereovision (three-dimensional vision) with enhanced depth perception. The middle ear of troödontids is also extremely enlarged, suggesting that they had an acute sense of hearing (*see* "What were the primary sensory capabilities of dinosaurs?," page 36). As we have warned against repeatedly, however, measuring quantities such as sensory capability from fossil remains is very treacherous, and these predictions can be considered only informed speculation, rather than testable hypothesis or empirically supported theory.

FIGURE 83. The skull of the small Mongolian theropod *Saurornithoides* differs from other birdlike dinosaurs in that the teeth have denticles or cusps, similar to those of some plant-eating dinosaurs. *Saurornithoides*, however, is presumed to have been a particularly agile predator, hunting by sight and grasping prey with its forelimbs.

SAURISCHIAN

Deinonychus antirrhopus

AMNH: *3015*
AGE: *Early Cretaceous, about 107 mya*
FORMATION: *Cloverly*
LOCALITY: *Buster Creek, Cashen Ranch, Crow Indian Reservation, Montana*
COLLECTOR: *Barnum Brown and Peter Kaisen*
DATE: *1931*

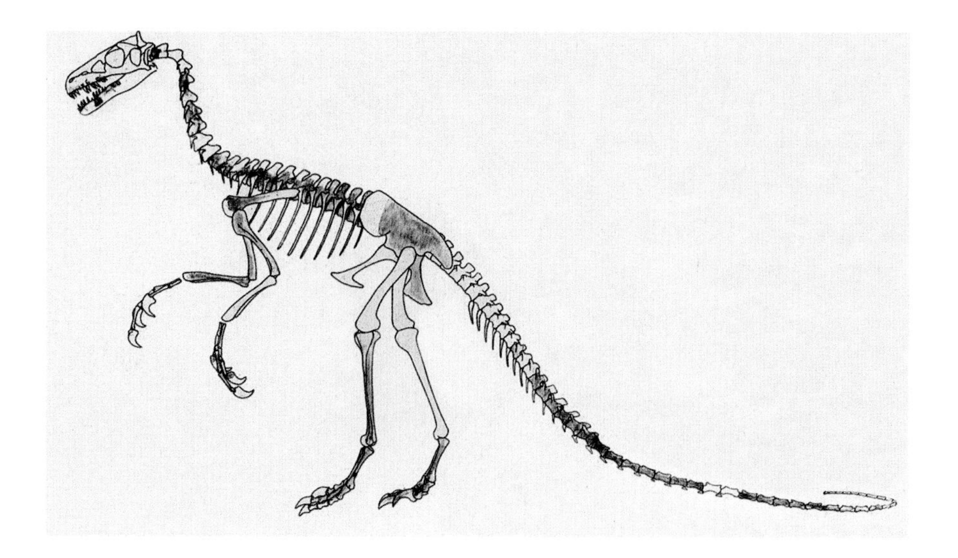

FIGURE 84. This reconstruction from the 1930s, made for Barnum Brown, demonstrates how well his *"Daptosaurus"* (later named *Deinonychus*) was known and how far his research on this animal had progressed.

Among dinosaurs discovered in the second half of the twentieth century, *Deinonychus* is one of the most spectacular and important. Study of this theropod changed the way we view dinosaurs and precipitated the reemergence of the idea that modern birds are a subgroup of theropod dinosaurs (*see* "Why are birds a type of dinosaur?," page 11).

Deinonychus was a small to medium-sized theropod, with a body about 2.5 meters long. Its thin, hollow bones indicate that it probably was an active, lightly built, running animal. Its stiff tail, supported by long bony rods, would have been an effective counterweight for the body when the animal ran at high speeds. Some researchers have suggested that *Deinonychus* was an intelligent, cunning, carnivore that hunted large herbivores, such as *Tenontosaurus*, in packs. Although the image is spectacular, unfortunately no evidence—beyond the common occurrence of more than one *Deinonychus* individual at a locality associated with a *Tenontosaurus* skeleton—has been brought to bear on this question. As always, other explanations are equally likely, such as chance association or predator traps (*see "Allosaurus,"* page 110), that could explain the occurrence of groups of individuals.

Deinonychus and many other maniraptors possess a specialized second toe that ends in an enlarged cutting claw. This toe probably was used to kill prey. Some fossil evidence of *Velociraptor* (a close relative) may support this idea (*see* "Did dinosaurs fight?," page 42). In addition, many modern carnivorous birds use enlarged foot claws for impaling and ripping apart their prey. This toe, combined with the large serrated teeth and the recurved claws on the hands, made *Deinonychus* a formidably armed predator.

Deinonychus is the most common theropod in the Early Cretaceous Cloverly Formation. American Museum of Natural History field parties, under the leadership of Barnum Brown, made several trips to these beds in southern Montana, amassing a large diversity of dinosaur fossils, including the skeleton (skull fragments, nearly complete limbs, and a series of vertebrae) of a small theropod (AMNH 3015) collected in 1931. Because small theropods were so rare, this specimen immediately piqued Brown's interest. Although he informally named the specimen *Daptosaurus*, and had a series of drawings produced, Brown never published his work. Consequently, John Ostrom's exploration of the Cloverly Formation in 1964 (*see "Microvenator,"* page 126), which produced additional specimens of this dinosaur, resulted in its official name, *Deinonychus antirrhopus*. Because the name *Daptosaurus* was never published and therefore is invalid (*see* "How do dinosaurs get their names?," page 8), the Museum has a specimen that is referred to as *Deinonychus,* rather than owning the type specimen of this animal.

The Museum's *Deinonychus* mount is the only one of this animal anywhere that includes real material. Because even this specimen (AMNH 3015) is incomplete, some bones in the mount are casts of specimens from Yale and Harvard. Other parts that are not known from any *Deinonychus* specimen were modeled from closely related animals, such as *Velociraptor*. The skull and pelvis of the Museum's mount incorporate modifications based on new research. During preparation for mounting, the Museum staff discovered that plans to mount this specimen had been initiated back in Brown's day; many of the foot and hand bones had been drilled to accept an armature (*see* "How are dinosaur mounts assembled in museums?," page 89). Because the plans were not carried out, we have had to wait almost 65 years for this small theropod to make its 1995 New York debut.

FIGURE 85. The large, highly modified sickle claw of *Deinonychus* gives it its name, meaning "terrible claw." Such claws are characteristic of *Deinonychus* and its close relatives, like *Velociraptor*.

SAURISCHIAN

Velociraptor mongoliensis

AMNH: *6515*
AGE: *Late Cretaceous, about 72 mya*
FORMATION: *Djadoctha*
LOCALITY: *Flaming Cliffs, Mongolia*
COLLECTOR: *Peter Kaisen*
DATE: *1923*

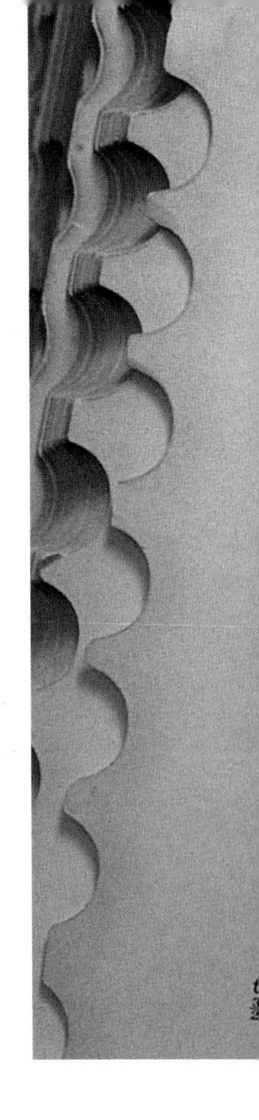

One of the more specialized hunters of the Late Cretaceous was *Velociraptor*, which is a member of the Dromaeosauridae, a group of theropods that shared a close common ancestor with birds. Except for being smaller, *Velociraptor* is extremely similar to *Deinonychus* (*see* page 129). Both possess tails with stiffening rods, long forelimbs with grasping hands, and a large raptorial claw on the second toe of the foot. *Velociraptor* also had a relatively large brain, large eyes, and may have had advanced sensory capabilities similar to those of its living bird relatives.

In North America the remains of dromaeosaurs are extremely rare. To date only four kinds have been described: *Saurornitholestes* and *Dromaeosaurus* from Alberta, *Deinonychus* from Montana, and *Utahrapto*r from Utah. By comparison, the record in Asia is spectacular. Although the material is not common, the diversity is remarkable, considering that much less collecting has been done on that continent than in North America. Several species in addition to *Velociraptor* have been discovered, including some new, undescribed forms that have not yet been named (*see* "How do dinosaur get their names?," page 8). Dromaeosaur diversity is so great that even at single localities more than one species is commonly present. Even giant forms, equivalent in size to the North American *Utahraptor*, have been recovered.

Many, if not most, of the *Velociraptor* remains collected by the joint field parties of the Mongolian Academy of Sciences and the American Museum of Natural History (*see* "The Western Gobi," page 209) show signs of predation or scavenging. Mongolia appears to have been a very dangerous place during the Late Cretaceous. In addition to flesh-eating dromaeosaurs, troödontids and carnosaurs also hunted. Why theropod dinosaur remains are so common in Mongolia but so rare in the rest of the world is difficult to understand, and the type of information that one could collect in the fossil record that would resolve this problem is hard to imagine.

FIGURE 86. *Velociraptor* was first described in 1924, shortly after it was collected at Mongolia's Flaming Cliffs in 1923. For over 50 years it was an obscure dinosaur, until additional remains were collected by Russian, Mongolian, Polish, and American expeditions. Since then it has become a movie star in *Jurassic Park*, as well as an important element in our understanding of the origin of birds.

The type specimen of *Velociraptor* was collected by Peter Kaisen, who accompanied the Central Asiatic Expedition on several campaigns. The specimen was described by Osborn in the same paper that contained his descriptions of *Oviraptor* and *Saurornithoides.* These were three of the most important, although at the time unappreciated, discoveries of these expeditions. In his 1924 description, Osborn described *Velociraptor* as an *"alert, swift-moving carnivorous dinosaur."*

Velociraptor has gained fame as the villain of Stephen Spielberg's 1993 film *Jurassic Park*. In the movie, *Velociraptor* was depicted as a large, hissing, cunning predator with a matriarchal pack structure and a vindictive disposition. In real life, the animals were much smaller, about 2 meters long; nevertheless, they were probably effective predators. We'll let the aficionados of action films speculate on family life, spitting, and hissing, because the fossil record provides no direct evidence that bears on this. Similarly, those who depict *Velociraptor* as a feathered "protobird" have no fossil evidence for this supposition.

SAURISCHIAN

Mononykus olecranus

AMNH: *28508*
AGE: *Late Cretaceous, about 72 mya*
FORMATION: *Nemegt*
LOCALITY: *Bogin Tsav, Mongolia*
COLLECTOR: *Namsarai B.*
DATE: *1987*

In the last several years the early history of birds has become increasingly well known. Many new Mesozoic species are being named and new specimens of previously known animals have been collected. One of the most unusual and unexpected is *Mononykus olecranus* from the Late Cretaceous of Mongolia.

Mononykus is a small animal, about the size of a turkey. It is gracile and long-legged with a sinuous S-shaped neck. *Mononykus* shares characteristics with modern birds, indicating that they shared a close common ancestor. Some of these features can easily be compared with those of a chicken. In *Archaeopteryx*, nonavian dinosaurs, and most other tetrapods (including humans), the small bone of the lower leg (the fibula) extends from the knee to the ankle. In *Mononykus* and other birds the fibula fails to reach the ankle, ending instead in an attenuated point. This is the small "toothpick" bone on the side of a chicken drumstick. The breastbone of *Mononykus* has a large keel, just like the chicken breastbone that forms the attachment point for much of the white meat. In nonavian dinosaurs the breast bone is flat, without a large keel. Many other features support the idea not only that *Mononykus* is a bird, but that it is more closely related to modern birds than is *Archaeopteryx*.

Mononykus, however, is a very unusual bird. It retains many primitive characters, evolutionarily lost in later birds, that bespeak its nonavian dinosaur heritage. Examples include teeth and a long tail. Its most unusual features, however, are its highly modified forelimbs. Instead of the elongate arms of other maniraptors, which evolved into the wings of birds, *Mononykus* has comical short stubs that sport single large claws. These appendages are not small, weak arms, but show the characteristic bumps and ridges found in digging animals with powerful forelimbs—such as moles and aardvarks. What did *Mononykus* use these strange arms for? We have no definite answers, but they were probably involved in some specialized feeding behavior.

FIGURE 87. ***Mononykus olecranus*** **was a very weird animal. The gracile legs, the small head with minute teeth, and the long tail all seem to suggest that it was a fast, specialized runner. But what could the front limbs be used for?**

Our quest to understand *Mononykus* has just begun. Only two years have passed since the first specimens were recognized, and parts of the skeleton are still incompletely known. More than ten new specimens (including at least three skulls) of a new animal very closely related to *Mononykus* have been discovered by American Museum of Natural History expeditions. This animal is called *Shuvuuia deserti*, and is sure to reveal more information about these unusual animals. Some answers may come from an unexpected source. Immediately after the initial specimen was recognized, a small skeleton labeled "birdlike dinosaur" from the 1923 Museum expedition to the Flaming Cliffs of Mongolia was discovered in the Museum's collection that has characteristics distinctive of *Mononykus*.

The Museum's mount of *Mononykus* is a cast based on a composite of two specimens: the type specimen, from Bogin Tsav, and a *Shuvuuia* specimen, found in 1993 at Ukhaa Tolgod. Information on the skull was garnered from a specimen collected during the 1994 field season.

size as those of a bird weighing a few kilograms. *Diatryma* itself had an estimated weight of well over 150 kilograms. It was not the largest known ground bird, however. That title goes to the extinct, 4 meter-tall, Late Pleistocene elephant bird (*Aepyornis maximus*) of Madagascar.

Traditionally, *Diatryma* was thought to be a cursorial (running) predator. Recent work has shown that the proportionally short and heavy leg bones made *Diatryma* an ill-suited runner. Instead, it probably walked in a slow, stately fashion, as do the cassowary and the bustard. The only well preserved skull (AMNH 6169) indicates that the end of the beak was straight as in leaf-eating birds, lacking the terminal hook found in avian carnivores. *Diatryma* also lacks the large hooked claws typical of archosaurian meat eaters.

Although there is no question that *Diatryma* is a bird, it does retain some primitive characteristics. These features, coupled with the gigantic size of the bones, explain why Osborn and Matthew initially identified some *Diatryma* foot bones as belonging to a nonavian theropod dinosaur. Recent analysis of *Diatryma* indicates that its closest relatives are Anseriformes, a group represented today by ducks and geese. Remains of *Diatryma* have been found in Eocene deposits of North America and Europe.

SAURISCHIAN

Presbyornis pervetus

AMNH: *28505*
AGE: *Early Eocene, about 50 mya*
FORMATION: *Green River*
LOCALITY: *Canyon Creek Butte, southern Sweetwater County, Wyoming*
COLLECTORS: *Storrs Olson, Allison Andors, Dan Chaney, and Per Erickson*
DATE: *1990*

Living birds comprise more than ten thousand species. The fossil record, especially in the Mesozoic and early Tertiary, is extremely poor. Fossils of these birds often can be referred only to groups with no living members, or they are isolated fragments that are hard to identify. An exception is the early Eocene bird *Presbyornis*, which is represented by many well-preserved skeletons.

FIGURE 90. Volcanic ashfalls sometimes produce mass burials that become bone beds. *Presbyornis*, a web-footed, storklike bird, was preserved in an ashfall in Wyoming 50 million years ago. *Presbyornis* may be related to the Anseriformes, a group including ducks. Here, a duck skull (below) is compared with the skull of *Presbyornis* (above).

Although detailed study of this animal has yet to be completed, preliminary study indicates that it was a primitive member of the Anseriformes, a group whose living members include ducks, swans, and geese.

In life *Presbyornis* stood about 45 centimeters tall. Its storklike, stilt legs reflect is shorebird habits. At some localities where *Presbyornis* is found, trackways showing its webbed feet are preserved. Like many shorebirds, it nested in colonies, as indicated by the presence of different size classes representing individuals of different ages found together in large concentrations at single sites. These concentrations may be due to catastrophic attrition of entire *Presbyornis* flocks. The Museum's specimens are thought to represent such a flock that was killed by a volcanic ashfall or associated storm. At other localities where *Presbyornis* has been collected the remains may not be the result of catastrophic occurrences. In large nesting colonies of shorebirds today, dead animals of several size classes litter the colonies.

Modifications to the skull and tongue bones of *Presbyornis* indicate that, like modern ducks, it had a suction filter-feeding apparatus. This feeding apparatus allowed *Presbyornis* to filter-feed on algae in the shallow, freshwater lakes of Eocene western North America.

SAURISCHIAN

Psilopterus australis

AMNH: *9157*
AGE: *Late Miocene about 21 mya*
FORMATION: *Canon de las Vacas*
LOCALITY: *Santa Cruz, Argentine Patagonia*
COLLECTOR: *Barnum Brown*
DATE: *1899*

P*silopterus* belongs to a group of predominantly South American ground birds called phorusrhacids. Some members of the group were as tall as 3 meters, with skulls up to 48 centimeters long. Most phorusrhacids had tiny wings that rendered them incapable of flight. Some of the smaller species, such as *Psilopterus*, however, had proportionally larger wings and may have been capable of extremely limited flight.

In contrast to the unrelated extinct giant ground birds of the Eocene (*see "Diatryma,"* page 137), phorusrhacids were fast runners, as their long legs (especially the lower part) and gracile skeletons indicate. Their long, hook-shaped beaks, similar to those of hawks or eagles, are a good indicator that these animals were highly carnivorous. Some workers have even suggested that phorusrhacids were the top carnivores in South America for most of the early Cenozoic. Only a few specimens of phorusrhacids have been collected outside South America. Most of these are from the Pliocene of Florida. This occurrence of phorusrhacids in North America is not surprising because many other animals with South American affinities (such as the armadillo-like glyptodonts) are also known from these beds. These occurrences suggest a faunal connection between North America and South America during the Late Pliocene.

The Museum's *Psilopterus* specimens were collected by Barnum Brown in Patagonia in 1899. As an example of the expectations Osborn had of Brown and his other employees regarding their personal commitment to the field, consider the following entry, dated December 7, 1898, in Brown's Patagonian journal: *"Yesterday about three hours before the* Casac *was to sail I was notified by Prof. Osborn that arrangements had been made for me to go to South America. Four of the Dept. men packed up my kit and I took another with me home to pack up. Imagine getting an outfit together in three hours to go on a seven thousand mile journey, to be gone a year or more. Such is the life of a fossil man."*

FIGURE 91. This beautiful illustration of a skull of the extinct, presumably carnivorous, flightless bird *Psilopterus*, was drawn by Museum artist Erwin Christman in 1899. The specimen was found by Barnum Brown in rocks of Miocene age in Argentina. In birds the hook-shaped beak is usually indicative of a meat-eating habit.

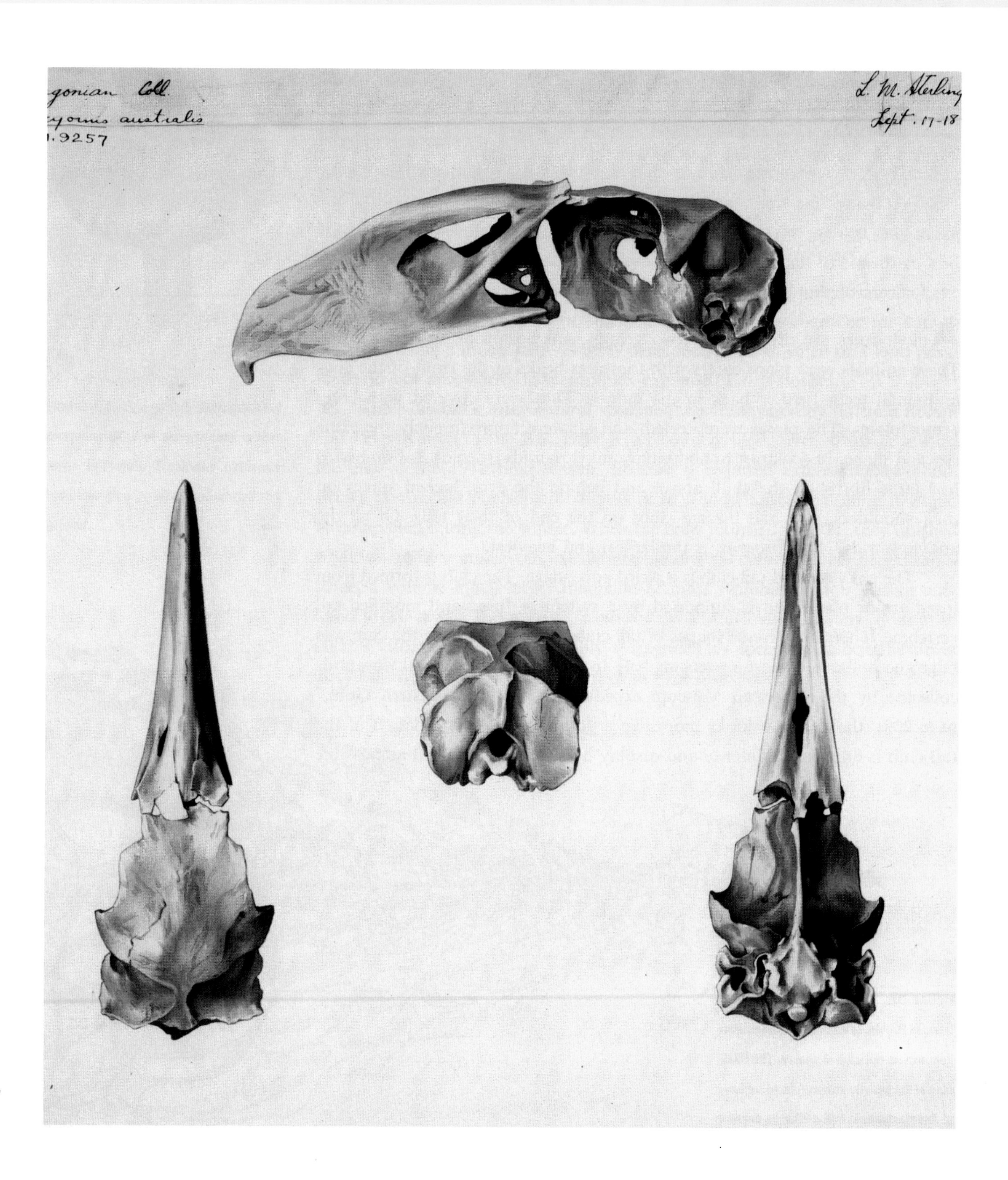
gonian Coll.
yornis australis
1.9257
L. M. Sterling
Sept. 17-18

a single species is that during the Late Cretaceous our planet's climate was much more uniform. This uniformity lessened the climatic influences of latitudinal bands, so plants and animals spread out into much wider (or perhaps longer) geographic ranges than they do today. This expansion was especially true for animals living along the shallow continental seaway that extended across North America from the Gulf of Mexico to up near the Arctic Ocean (*see* "Why did nonavian dinosaurs become extinct?," page 62).

ORNITHISCHIAN

Stegosaurus stenops

AMNH: *650*
AGE: *Late Jurassic, about 140 mya*
FORMATION: *Morrison*
LOCALITY: *Como Bluff region, Wyoming*
COLLECTOR: *P. C. Kaisen*
DATE: *1901*

S*tegosaurus* has been an enigma since its first remains were recovered more than a century ago. Its relationships are not the problem. Everyone agrees that *Stegosaurus* and its plated relatives, such as *Dacentrurus* from Great Britain, *Kentrosaurus* from Africa, and *Huayangosaurus* from China, are ornithischians, belonging to a group called Stegosauria that arose from a common ancestor with plates and/or spikes. Stegosaurs are, in turn, closely related to the other group of armored dinosaurs, the ankylosaurs. Together, they form a group called Thyreophora. Rather, the debate has centered on two specific aspects of stegosaur anatomy: the placement and function of the body armor, and the presence of the "second brain" in the pelvis.

Stegosaurs have an enlarged area in the spinal canal called the sacral plexus. In life this cavity housed the spinal ganglia in the sacrum (vertebrae that are fused to the pelvis). Other dinosaurs (including living ostriches) possess this enlargement; however, in none is it developed to the degree that it is in *Stegosaurus*.

FIGURE **96.** **Probably the most distinctive dinosaur, *Stegosaurus* has a double row of plates down the back and spikes on the tail. Despite its popularity, there are less than a dozen good specimens of *Stegosaurus*, and it is not common at any locality.**

When *Stegosaurus* was first discovered, analysis of its cranial cavity showed that it had a tiny brain. The adult *Stegosaurus* brain weighs only about 75 grams and is no bigger than a walnut. The small brain is in stark contrast to the large adult body size. How could such a small brain be the control center for complex movement and behavior in such a lumbering brute?

One idea was that the sacral plexus formed a "second brain" that controlled the back half of the body. This idea is interesting but hard to test. Although many of the extinct dinosaurs reached giant sizes (*see* "How large were the biggest dinosaurs?," page 22), the fossil evidence (especially from fossil trackways) shows that they were very capable of living active lifestyles and exhibiting complex behaviors.

Another interpretation is that the enlarged sacral plexus functioned as a complicated switching center for neural control of muscle coordination, as it does in ostriches. In this sense the sacral plexus is not a brain at all, in that it does not "think" or "remember." Barnum Brown proposed such a hypothesis in 1932. Stegosaurs, however, were slow-moving animals, and it is unlikely that they required complex switching for their hind limbs. Finally, if other dinosaurs are examined, many can be shown to have an enlarged sacral plexus. This enlargement corresponds to the position of the glycogen storing organ in living birds. It is likely that nonavian dinosaurs shared the presence of a glycogen body with birds. The function of the glycogen body is unknown.

The body armor is the most familiar feature of stegosaurs, whose name means "spiked lizard." Different kinds of stegosaurs have dramatically different arrangements of spikes and plates. The most familiar arrangement is that of *Stegosaurus*, in which the plates are aligned along the back and the spikes are restricted to the tail. Other stegosaurs, such as *Dacentrurus*, had only spikes, while *Huayangosaurus* had both spikes and plates on its back.

Since *Stegosaurus* was described by O. C. Marsh in 1877, the arrangement of plates and spikes on its back and tail has been the subject of some controversy. Early *Stegosaurus* reconstructions configured the plates in various ways—as single or double, alternate or paired rows along the midline of the back, erect or hung limply over the sides of the animal.

Discovery during the 1980s of a new specimen of *Stegosaurus* in Colorado with a well-preserved set of plates has demonstrated that the plates were arranged in a single row. The plates do not lie exactly on the midline and because they alternate, plates are either left or right sided. Around the throat of this skeleton and others (including the American Museum of Natural History specimen), are hundreds of small bony ossicles, each about the size of a large button. These bones may have formed a sort of chain mail for protecting this highly vulnerable area.

The tail of all stegosaurs is armed with large spikes. In *Stegosaurus* these spikes are heavy, and can reach a length of 1.2 meters. The spiked tail may have served as a powerful offensive or defensive weapon, or perhaps its primary application was in display, either threatening or social. The only thing that we can say for sure about the spiked tail of *Stegosaurus* is that it must have imparted a de Sadian dimension to mating behavior.

ORNITHISCHIAN

Tenontosaurus tilletti

AMNH: *3034*
AGE: *Early Cretaceous, about 107 mya*
FORMATION: *Cloverly*
LOCALITY: *Mott Creek, Crow Indian Reservation, Montana*
COLLECTOR: *Barnum Brown and Peter Kaisen*
DATE: *1932*

FIGURE 97. A primitive relative of the hadrosaurs, or duck-billed dinosaurs, the Early Cretaceous *Tenontosaurus* had a less specialized dentition and lacked the extensive nasal expansions of hadrosaurs. During his extensive field operations in the Cloverly Formation of Wyoming, Barnum Brown discovered complete skeletons, as well as juveniles, of *Tenontosaurus*.

T*enontosaurus* is closely related to hadrosaurs, as well as to *Iguanodon*, *Camptosaurus*, and other ornithopods; it is the most primitive member of this group. Like its close relatives, *Tenontosaurus* inherited large nasal openings in the skull from a common ancestor with more advanced ornithopods, had a system of interconnected bony tendons that stiffened the backbone and tail, and was probably predominantly bipedal, although it occasionally walked around on all four limbs. Unfortunately, in the American Museum of Natural History's display specimen, many of the tendons, especially in the tail, were removed during preparation. Consequently, the mount is inaccurate; the tail should have no sinuous curve and should be held straight behind the animal, parallel to the ground.

In 1927 Barnum Brown traversed south-central Montana, a region that he had originally visited in 1903, at the advent of the Hell Creek work (*see* "The Hell Creek Beds," page 189). Brown reported to Walter Granger near the end of the season that he had located *"a fine little dinosaur about the size of* Camptosaurus.*"* This little dinosaur indeed turned out to be a fine skeleton—the first specimen of *Tenontosaurus* recognized. Later, after continued study of the material, it was discovered that Brown had collected a few *Tenontosaurus* vertebrae (AMNH 5854) during his earlier 1903 trip to the region.

Not until 1931 was Barnum Brown able to return to what would become one of his favorite hunting grounds, the Cloverly badlands on the Crow Indian Reservation. *Tenontosaurus tilletti* is the most abundant

Cloverly dinosaur and represents a primitive member of a group called the Iguanodontia. Brown recognized the little skeleton immediately as something of great scientific importance. In addition to *Tenontosaurus*, Brown and Kaisen found an entirely new and undescribed dinosaur fauna, including *Deinonychus, Microvenato*r, and *Sauropelta.* Although he began manuscripts on all of these animals, none were ever finished, and all waited to be formally named until John Ostrom picked up Brown's work in the 1960s (*see "Microvenator,"* page 126).

The American Museum of Natural History's specimen is a full mount that was constructed by Charles Lang and unveiled in 1938. Because the skull of this specimen was severely crushed, a plaster replica was added. This skull looked rather odd and lacked any detail beyond the fluted teeth. During the most recent renovation, a new skull was sculpted on the basis of material recently collected by the Museum of the Rockies in Bozeman, Montana.

ORNITHISCHIAN

Camptosaurus dispar

AMNH: *6120*
AGE: *Late Jurassic, about 140 mya*
FORMAION: *Morrison*
LOCALITY: *Bone Cabin Quarry, Wyoming*
COLLECTOR: *Peter Kaisen*
DATE: *1905*

C*amptosaurus* is a primitive ornithischian dinosaur from the Late Jurassic. Most *Camptosaurus* specimens have been collected in North America; one specimen of a distinct species, *Camptosaurus prestwichii*, is known from Great Britain. The American Museum of Natural History specimen, collected in 1905 at Bone Cabin Quarry, during the final season of the Museum's work there, is about 2 meters long; this is small by *Camptosaurus* standards (some specimens reach more than 5 meters). It was found in a semi-articulated position, indicating that it was buried before the skeleton was disturbed by scavengers or by the movement of water and sand.

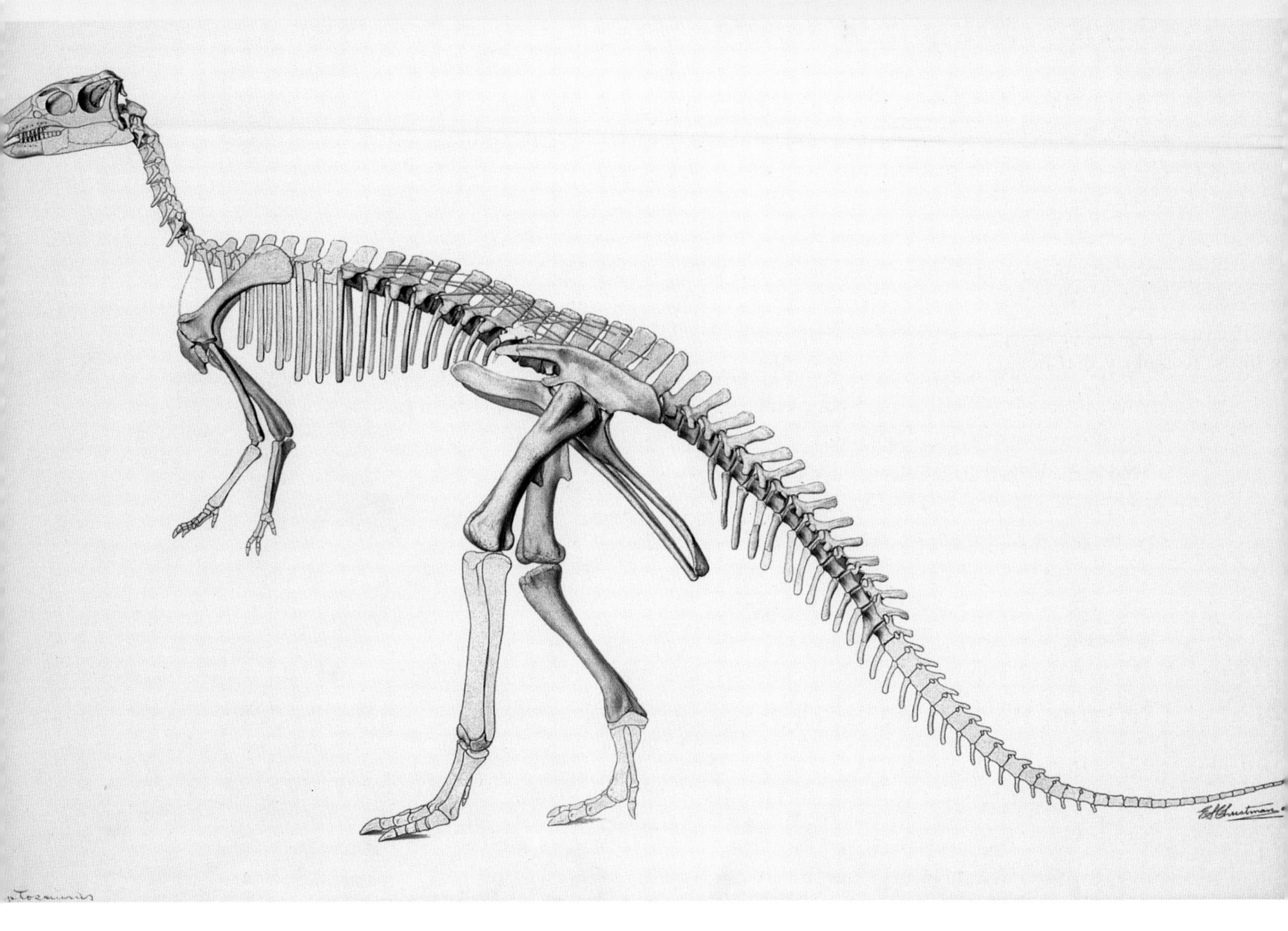

FIGURE 98. This beautiful drawing of *Camptosaurus*, the work of Erwin Christman, represents a small specimen collected at Bone Cabin Quarry. This specimen was one of the first small ornithischians displayed at the Museum.

Remains of *Camptosaurus* are relatively rare, and complete skeletons are unknown. The Museum's specimen was prepared for exhibit in 1908, and this specimen preserves a latticework of ossified tendons on the vertebral spines of the lower back. Many dinosaur specimens, especially ornithopods, show similar structures. In most cases these structures continue on the tail vertebrae. These tendons helped keep the back rigid and the tail stiff and parallel to the ground. Functional analyses, bolstered by footprint evidence, indicate that these dinosaurs were not the bipedal erect forms, as illustrated in the early twentieth century, but rather were predominantly quadrupedal, only occasionally rearing on two legs.

ORNITHISCHIAN

Prosaurolophus maximus

AMNH: *5386*
AGE: *Late Cretaceous, about 72 mya*
FORMATION: *Judith River*
LOCALITY: *Red Deer River region, Alberta, Canada*
COLLECTOR: *Barnum Brown*
DATE: *1915*

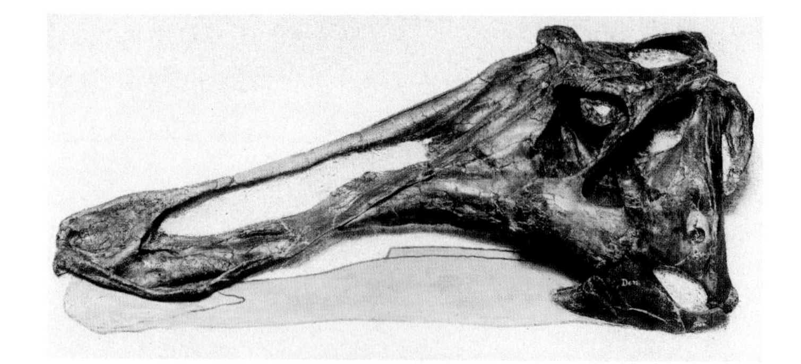

FIGURE 99. ***Prosaurolophus*** **differs from** ***Saurolophus*** **in that it has a smaller protrusion on the head. Unlike that of its close relative, this bony protuberance was solid rather than hollow.** ***Prosaurolophus*** **is part of the diverse hadrosaur fauna that occupied North America during the Late Cretaceous.**

Hadrosaurs are a diverse and cosmopolitan group of predominantly quadrupedal herbivores whose remains are ubiquitous in Late Cretaceous deposits. Hadrosaurs are popularly called duck-billed dinosaurs because of their toothless bills. Unlike ducks and other advanced avians, however, hadrosaurs had teeth. These teeth were arranged in tooth batteries—a dental mosaic consisting of thousands of teeth compressed into ever-growing masses. This battery was set back in the jaw well above the jaw joint in a mechanically advantageous position for exerting maximum force. The tooth batteries of hadrosaurs constantly produced a rough crushing surface to mash plant material cut off by the slicing beak.

Evidence from trackways suggests that hadrosaurs were quadrupedal, carrying their spines nearly parallel to the ground. The tail was stiff and extended backward, held high off the ground as a counterweight for the trunk. This image is in contrast to older reconstructions, which portrayed duckbills as bipedal. No doubt this would have come as quite a shock to Charles Sternberg, who in 1912 had very different ideas about the posture and habits of hadrosaurs: *"The duck-bill lived in the bayous of the country. He was a powerful swimmer, and could use his great hind limbs, eight feet long, in the same manner as a frog uses his. Or, while feeding on the rushes that lined the sluggish streams, he could plant his powerful hind feet in the sandy bottom, while, with his front ones acting as arms, he could pull into his duck-billed mouth the succulent forage."*

Prosaurolophus is a nondescript dinosaur that is poorly studied. It is closely related to *Saurolophus*, and the two are very similar, except that *Prosaurolophus* is larger and has only a small bump on the back of its skull where *Saurolophus* has a pronounced crest.

ORNITHISCHIAN

Saurolophus osborni

AMNH: *5220*

AGE: *Late Cretaceous, about 72 mya*

FORMATION: *Horseshoe Canyon*

LOCALITY: *Tollman Ferry, Red Deer River region, Alberta, Canada*

COLLECTOR: *Barnum Brown and Peter Kaisen*

DATE: *1911*

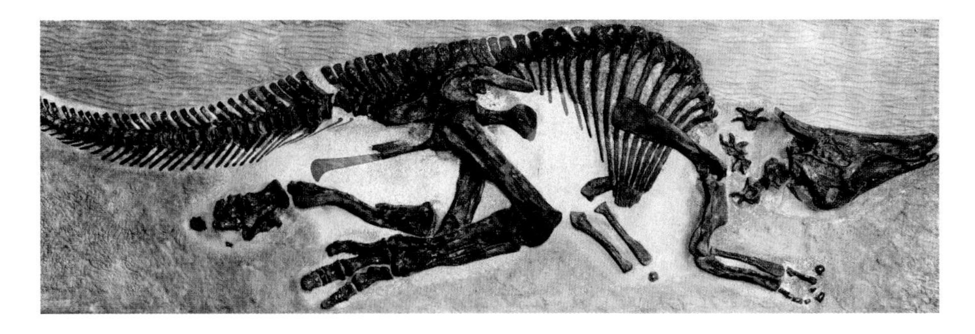

FIGURE 100. This nearly complete skeleton of the Late Cretaceous hadrosaur *Saurolophus* was found lying on sand with ripple marks and other signs of rapid burial in a stream or river. Did *Saurolophus* actually live in the river or was it an upland animal that died while trying to cross the river? We would like to know the answers to such questions, but for extinct animals they must remain elusive. Barnum Brown collected this superb skeleton during his work in the Red Deer River region of Canada in what is now Dinosaur Provincial Park.

The duck-billed dinosaur *Saurolophus* has a short spike instead of the large crest that is typical of lambeosaurines like *Corythosaurus* (see page 158). In life the spike may have supported a cartilaginous expansion of the nose. The type specimen of *Saurolophus osborni* (AMNH 5220) is a well-preserved skeleton found by Barnum Brown along the Red Deer River. *Saurolophus* remains have also been found in Montana and, interestingly, in Mongolia. The discovery of the Mongolian species (*Saurolophus angustirostris*), along with occurrences of other closely related dinosaurs, such as ornithomimids and tyrannosaurs, has prompted speculation that North America and Asia were connected by a land bridge during this part of the Cretaceous.

This specimen was collected in 1911 on a sandy surface covered with ripple marks, indicating that the specimen had come to rest on the bottom of a lake or stream. On several parts of the skeleton patches of skin are preserved. The skeleton was mounted during the winters of 1912 and 1913. Charles Falkenbach and George Sternberg prepared the specimen in relief by chiseling away the rock but leaving most of the bones attached to the matrix (*see* "How are dinosaur mounts assembled in museums?," page 89). Thus, the skeleton is preserved exactly as it was found in the rock. The right side, upon which the skeleton had rested, was chosen for display, because it had been protected from weathering. The bones were crushed somewhat from the weight of the rock deposited on them. Missing elements were either left as gaps (the neck and left hand) or were painted on the plaque (the vertebrae).

ORNITHISCHIAN

Edmontosaurus annectens

AMNH: *5060*

AGE: *Late Cretaceous, about 65 mya*

FORMATION: *Lance*

LOCALITY: *Converse County, Wyoming*

COLLECTOR: *Sternberg family*

DATE: *1908*

The Sternbergs (*see* "The Red Deer River," page 203) were working under contract to the British Museum in the Lance Creek area of Wyoming, near the town of Lusk in the summer of 1908. Their charge was to secure ceratopsian specimens like *Triceratops* for the British Museum collection. The work was hard, and the results had not been particularly rewarding. Just when supplies had nearly run out and the situation was looking desperate, Charles Sternberg and his three sons uncovered what was to become one of the world's greatest paleontologic treasures: the *Edmontosaurus* mummy.

FIGURE 101. The Sternberg family discovered the *Edmontosaurus* "mummy" in the Late Cretaceous deposits around Lance Creek, Wyoming, in 1908. The extent of skin imprint preserved is unusual, but much care was taken with the collection of this specimen.

Although the specimen was most likely destined for the British Museum, when Henry Fairfield Osborn got wind of the discovery, he used his influence to secure the mummy for the American Museum of Natural History. In 1908 the museum purchased the mummy for the then gigantic sum of $2,000. No doubt this hefty fee influenced Charles Sternberg's words in 1908 that, *"I am glad the American Museum, by far the noblest in America and soon to be the greatest on earth, under the splendid management of Prof. Henry Fairfield Osborn, is to be the final resting place of this great lizard."*

When the skeleton was found, the tail, hind feet, and the hind portion of the pelvis had already been exposed and had eroded away. Presumably these missing portions were originally preserved. The rest of the animal is mounted in the Museum as it was discovered; no regions were restored, and all the remaining bones are connected. The dinosaur lies on its back, with its knees drawn up, its chest open and upturned, its forelimbs outspread, and its neck

and head twisted sharply downward and backward toward the right side. The hands of this and other specimens were originally thought to indicate webbing, as might be predicted for an animal thought to live in an aquatic habitat. This observation is inconsistent, however, with the footprint evidence, which suggests that hadrosaur forefeet possessed a hard, horseshoe-shaped pad. Further study has shown that the webbing is due to postmortem displacement of the loose skin on the hand.

In 1912 Henry Fairfield Osborn described AMNH 5060 in detail and advanced the following theory of how the specimen came to be preserved:

> *After a natural death (in other words, not death by predators) the body lay exposed to the sun for a long time, perhaps on a sand bar or in a stream. The muscles and soft internal tissues became completely dried and shrunken while the skin, hardened and leathery, shrank around the limbs and was drawn down along the bones. In the stomach and abdominal areas the skin was drawn within the body cavity, while along the sides of the body and on the arms, it was formed into creases and folds. At some later date, the "mummy" may have been caught in a sudden flood and carried downstream and rapidly buried in fine sand and clay. A cast, or impression, of the skin formed in the sand before the skin and other soft parts decayed. There is no remnant of the actual skin preserved only its imprint.*

FIGURE 102. The *Edmontosaurus* mummy was found lying on its back with its head pulled under the body and its right arm thrust into the air. The skin of the chest and abdomen was shrunken and collapsed onto the bones. The hind limbs and tail were lost to weathering before its discovery.

The Sternbergs collected a second *Edmontosaurus* specimen from the same site. This specimen, which was not as well preserved as AMNH 5060, they sold to the Senckenberg Museum in Germany.

ORNITHISCHIAN

Anatotitan copei

AMNH: *5730, 5886*
AGE: *Late Cretaceous, about 65 mya*
FORMATION: *Hell Creek*
LOCALITY: *Moreau River, South Dakota, and Crooked Creek, Montana*
COLLECTOR: *J. I. Wortman and R. S. Hill (for E. D. Cope); Barnum Brown*
DATE: *1882, 1906*

A*natotitan* is a medium-sized hadrosaur. Cursory examination of this dinosaur's skull reveals why hadrosaurs are commonly called duckbills. In the American Museum of Natural History's Hall of Ornithischian Dinosaurs, two *Anatotitan* specimens were mounted together, in an effort to convey an image of life in the Cretaceous. Since these specimens were mounted, we have reinterpreted the posture and locomotory habits of hadrosaurs. Instead of standing upright or crouching low with sagging tails, our new view of hadrosaurs is that they were quadrupedal, with stiff, outstretched tails held parallel to the ground (*see "Prosaurolophus,"* page 152).

The mounted skeletons have an interesting history. One of the skeletons (AMNH 5730) was purchased, unmounted, in 1899 as part of the Cope collection. Unlike the *Allosaurus* skeleton (see page 110), which Cope had never unwrapped, Cope had studied and described this skeleton. It was found north of the Black Hills in South Dakota in 1882 by two of Cope's collectors, R. S. Hill and J. I. Wortman (who later joined the staff of the American Museum of Natural History).

The other mounted specimen (AMNH 5886) came from Crooked Creek in central Montana and was found by a rancher, Oscar Hunter, while riding through the Badlands with a companion in 1904. The specimen was partly exposed with its backbone and ribs sticking out above the ground. The two men argued about the nature of the skeleton. To settle the dispute, Hunter dismounted and kicked the tops off the vertebrae, proving by their brittle nature that they were stone rather than recent bison bones. As Barnum Brown related: *"The proof was certainly conclusive, but it was extremely exasperating to the subsequent collectors."* Alfred Sensiba, another cowboy, heard of the skeleton and bought it from Hunter for a pistol. Brown in turn purchased it from Sensiba for $250 and excavated it in 1906.

FIGURE 103. This specimen was collected for E. D. Cope in 1882. The photo dates from Brown's later years, when he was the unchallenged "king of the dinosaurs."

FIGURE 104 (OPPOSITE). The skull of the Cope specimen of *Anatotitan* was the subject of a beautiful set of drawings by the unsurpassed scientific illustrator Erwin Christman. The gigantic nasal opening, the dental battery with hundreds of teeth, and the toothless beak convey both aesthetic and scientific information.

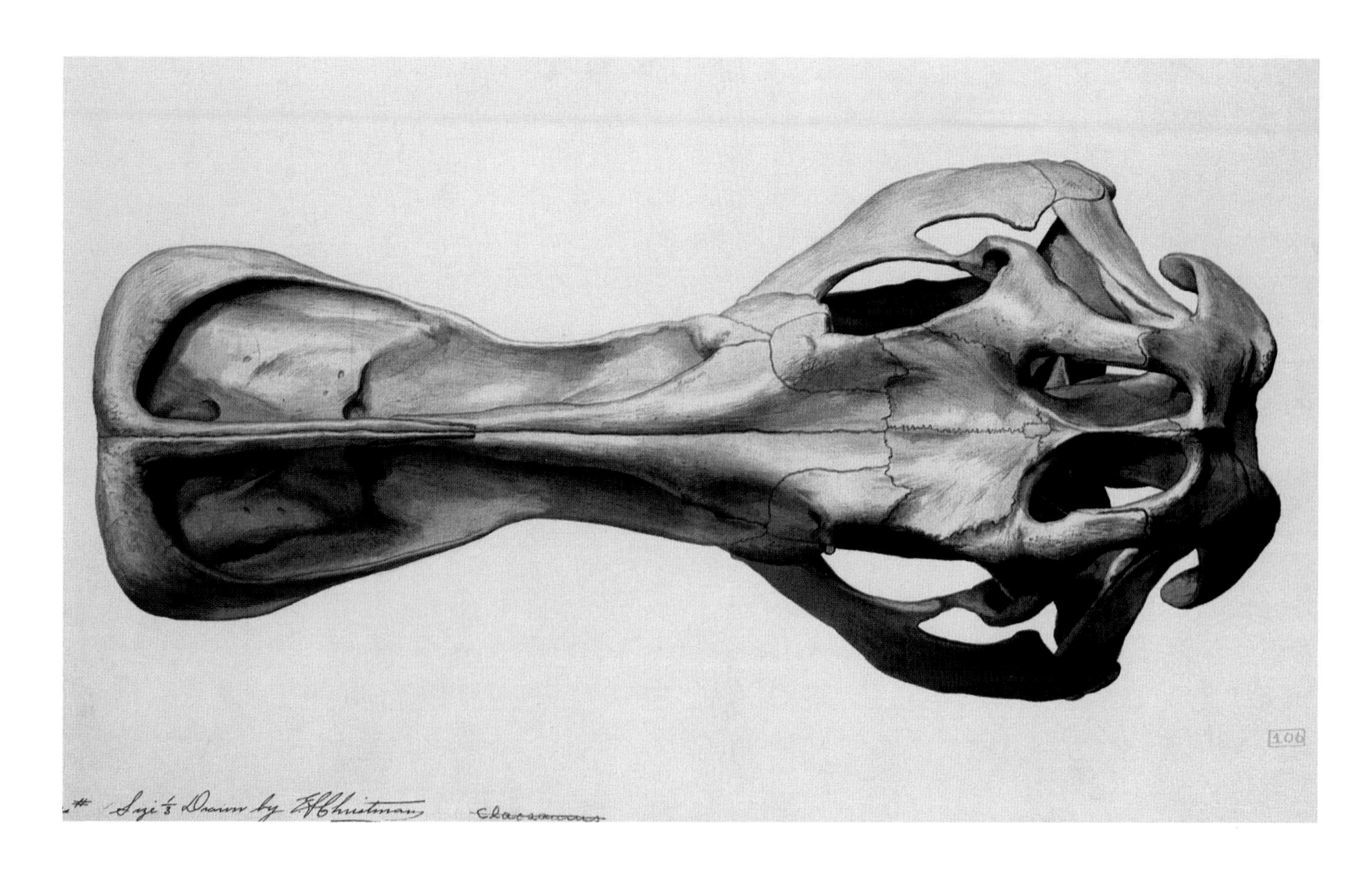
106
Size ⅓ Drawn by E. Christman

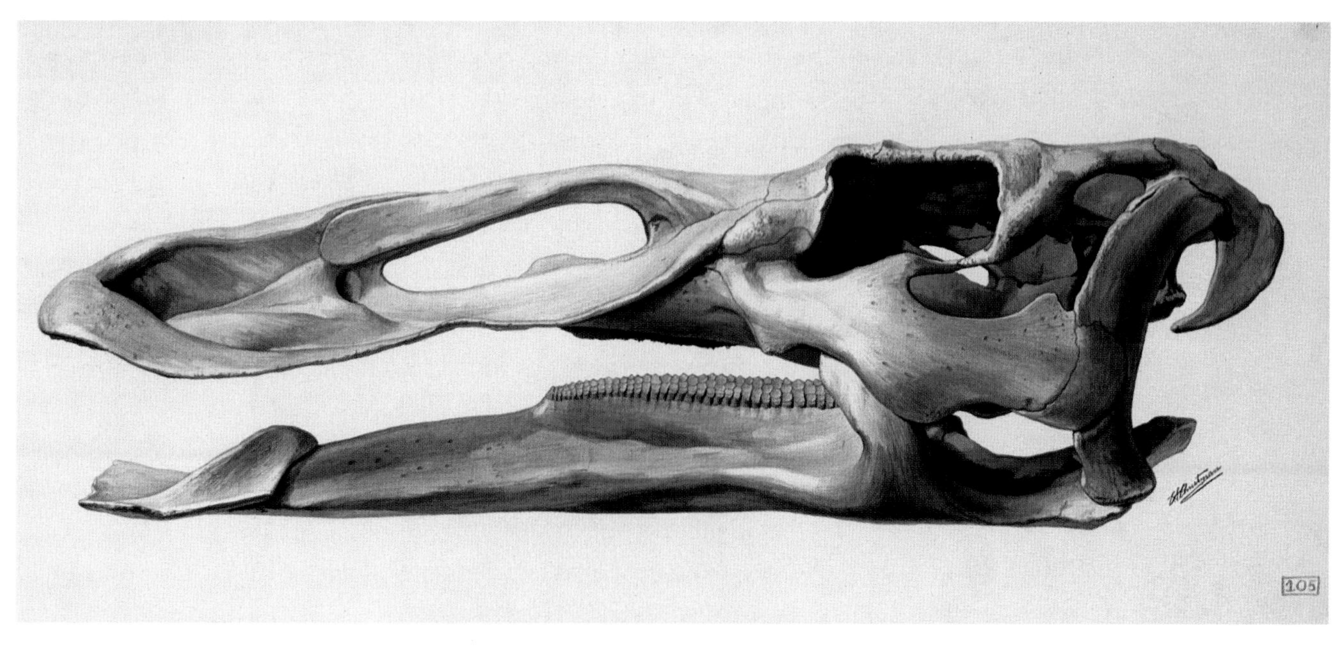
105

For more than 75 years the Museum's skeletons were thought to be representatives of *Anatosaurus annectens*. Further study of hadrosaur relationships, however, indicated that the name *Anatosaurus* was invalid (*see* "How do dinosaurs get their names?," page 8), so a new name had to be chosen. In 1990 the name *Anatotitan* (Greek *titan*, "large" + Latin *anas*, "duck") was proposed for these specimens.

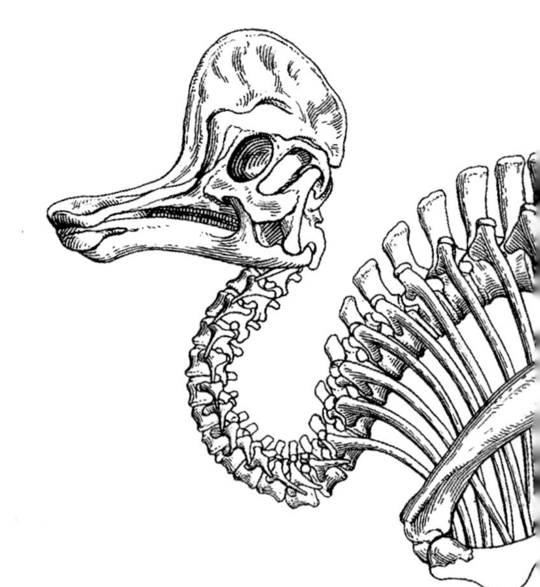

ORNITHISCHIAN

Corythosaurus casuarius

AMNH: *5240, 5338*
AGE: *Late Cretaceous, about 72 mya*
FORMATION: *Judith River*
LOCALITY: *Steveville and Sand Creek, Red Deer River region, Alberta, Canada*
COLLECTOR: *Barnum Brown and Peter Kaisen*
DATE: *1912, 1914*

FIGURE 105. *Corythosaurus* is a Late Cretaceous hadrosaur characterized by a very large, hollow expansion on the top of the skull. First discovered by Barnum Brown in 1912, *Corythosaurus* was soon represented by some superb skeletons, as well as skin imprints. Although all the *Corythosaurus* skeletons were preserved by rapid burial in rivers, their mode of life and the function of the nasal enlargements are still disputed.

All lambeosaurines have large, hollow, elaborate crests on their heads that are characteristic of each species and may have functioned in vocalization or display (*see* "Did dinosaurs make sounds?," page 34). The half moon-shaped crest of *Corythosaurus* gives this dinosaur its name, which means "helmeted lizard."

A complete skeleton of *Corythosaurus* (AMNH 5240) was one of the finest dinosaur skeletons found by Barnum Brown in the Red Deer River country of Alberta. Discovered in 1912, the skeleton is complete except for parts of the forelimbs and tail, and is remarkable for its many skin impressions, as described by Brown:

"The impression of the epidermis covering the greater part of the body has been skillfully worked out in detail by Mr. Otto Falkenbach and although faint in places, where covered by masses of vegetable material, the general pattern is fairly well determined, likewise the outline of the neck, body and tail. It is evident from attendant circumstances of deposition that the carcass had drifted on a beach. The bedding plane under the body was unusually irregular; a complete skeleton of a young Baenid turtle, No. 5421, and several water-worn Trachodont bones were lying under the tail."

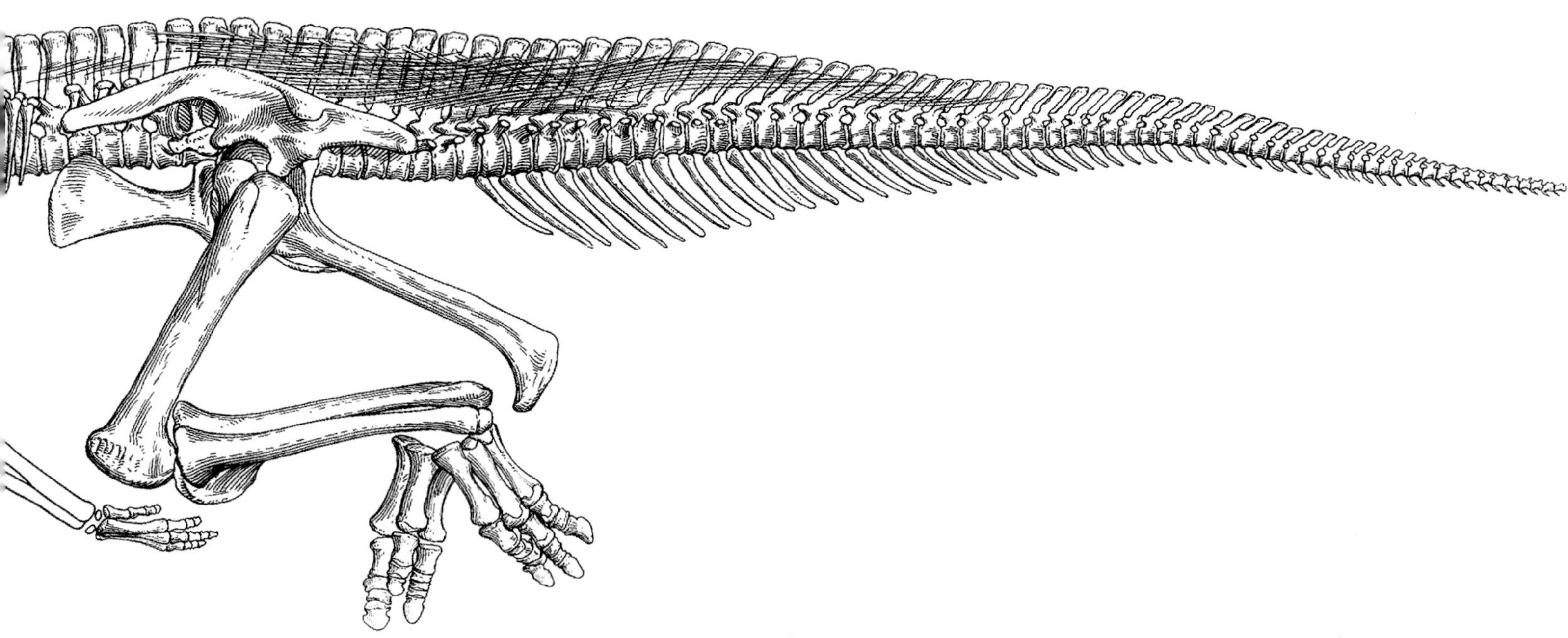

Brown and Kaisen found another superb skeleton of *Corythosaurus* (AMNH 5338) along the Red Deer River in 1914. This skeleton lacks much of the tail (missing skeletal parts were painted in for display), but it has good forelimbs and one of the most complete and perfectly preserved skulls known among dinosaurs. The thin rodlike elements crisscrossing the vertebrae above the pelvis are ossified tendons, common in hadrosaurs and apparently helping to make the backbone rigid. Both of the *Corythosaurus* specimens on display at the American Museum of Natural History lie in their original death poses.

Also in the Ornithischian Dinosaur Hall at the Museum is another beautifully preserved hadrosaur specimen (AMNH 5340). In 1920 William D. Matthew described this animal as a new, small type of hadrosaur and named it *Procheneosaurus*. It now seems likely that this animal is a juvenile of one of the common Judith River hadrosaurs, *Corythosaurus* or *Lambeosaurus*. Why can't we tell which one? Most of the characters used to differentiate these animals did not develop until late in life. Many modern animals exhibit the same phenomenon: The young of closely related forms are very similar, even though the adults have disparate body structures. Complete growth series of these animals are very rare, and only the collection of additional specimens or the discovery of new diagnostic features will allow us to determine which species *Procheneosaurus* belongs to.

ORNITHISCHIAN

Hypacrosaurus altispinatus

AMNH: *5278*
AGE: *Late Cretaceous, about 70 mya*
FORMATION: *Horseshoe Canyon*
LOCALITY: *Red Deer River region, Alberta, Canada*
COLLECTOR: *P. A. Bungert and George Sternberg*
DATE: *1915*

Like other lambeosaurine hadrosaurs, *Hypacrosaurus* was a large quadrupedal herbivore. It is distinguished from its close relatives by the very tall spines on its vertebrae and the shape of the crest on its skull. The nearly complete specimen on display at the American Museum of Natural History was found by P. A. Bungert and collected by George Sternberg in 1915 for the Canadian National Museum in Ottawa. Through exchange it was acquired by the Museum in 1925.

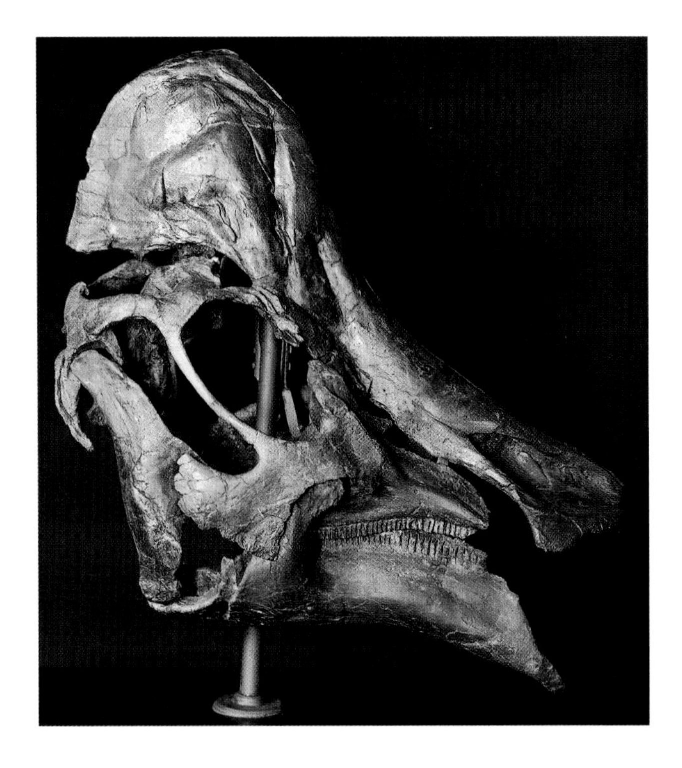

FIGURE 106. Hadrosaurs are extremely difficult to tell apart. Differences in head ornamentation, such as the large hollow crest in this *Hypacrosaurus*, are the only features that differentiate some species. The same is true of modern African antelopes, among which many similar species differ only in the shape of their horns.

Nearly complete growth series of *Hypacrosaurus* have been collected and identified by field parties from the Museum of the Rockies in Bozeman, Montana. Juvenile specimens and hatchlings are both on display at the Museum. Although associating juvenile specimens with adults is tricky and mistakes are common, the enlarged spines on the *Hypacrosaurus* vertebrae make identification in this case definitive.

Representative growth series provide us important information regarding changes in shape during growth. These changes are the result of different parts of the body growing at different rates as an organism ages (a phenomenon called allometry). Familiar examples of body parts affected by allometry include the disproportionately large feet of puppies and the large heads of human babies. As puppies mature, their bodies grow at a faster rate than their feet. As human mature, their bodies grow more quickly than their heads. Thus, in adults these features are smaller relative to the rest of the body than they are in juveniles.

In dinosaurs, specifically lambeosaurine hadrosaurs, growth series allow us to determine several modifications. First, as in modern birds, dinosaurs grow into their eyes, in the same way that puppies grow into their feet. This change is accompanied by several other changes in the skull. Juveniles have a very short skull with no crest, and the dental battery (*see* "What kind of teeth do dinosaurs have?," page 30) is poorly developed. Older animals show greater development of these features, until finally in adulthood the crest is highly developed, the face is longer, and the dental battery is fully formed.

ORNITHISCHIAN

Lambeosaurus lambei

AMNH: *5353, 5373*
AGE: *Late Cretaceous, about 72 mya*
FORMATION: *Judith River*
LOCALITY: *Red Deer River region, Alberta, Canada*
COLLECTOR: *Barnum Brown and Peter Kaisen; Barnum Brown and Albert Johnson*
DATE: *1913, 1915*

Lambeosaurus is one of the crested lambeosaurine hadrosaurs with a bizarre head shape. Like its close relative *Corythosaurus*, the bony crest of the skull is hollow. As with other modifications to the skull, paleontologists have speculated that this chamber functioned as a resonating device for vocalization. Another theory is that the extensive head ornamentation was used to support soft-tissue structures, such as a frill, and *Lambeosaurus* and its relatives are often restored as such. Although this function has not been conclusively demonstrated, the fact that several mummies of closely related individuals (*see "Corythosaurus,"* page 158, and *"Edmontosaurus,"* page 154) have been collected makes it at least conceivable that a mummified specimen of *Lambeosaurus* with preserved soft tissue will turn up, enabling us to test this idea by direct observation. Enough specimens of *Lambeosaurus* have been collected to establish that, like the antlers of deer, the crests grew and became more elaborate later in life.

In the beds where *Lambeosaurus* is found, hadrosaur diversity is extremely high. *Lambeosaurus* shared its environment with at least nine other

FIGURE 107. One of the most spectacular hadrosaurs is this double-crested form, *Lambeosaurus*, from the Red Deer River. It is named after Charles Lambe, one of the pioneers of Canadian paleontology.

kinds of hadrosaurs (both lambeosaurines and hadrosaurines). Most were about the same size and similar in build. The major difference among them (especially the lambeosaurines) is the shape of their crests. This variation in crests has led some paleontologists to suggest that the crests operated as a species recognition feature, perhaps even involved in reproductive display. This same sort of argument has been used to explain the different shapes and sizes of horns in today's African antelopes. As were the Cretaceous floodplains of western North America, the African savannas are populated by animals that are very similar in most aspects of body plan but show a diverse spectrum of horn morphologies. Some data suggests that antelopes are drawn to members of their own species on the basis of horn size and shape. This argument has even been extended to the proposal that this sort of interaction (termed sexual selection) is the evolutionary mechanism behind the origins of diverse and unique horn shapes. Whether the head ornamentation of the Late Cretaceous hadrosaurs was subject to evolutionary forces comparable to modern antelopes cannot be ascertained empirically; we know only that there were many different kinds.

The Museum has two *Lambeosaurus* specimens. AMNH 5353, a skull, was collected during the 1913 field season. The skull was beginning to erode out of a steep bluff, nearly 50 meters high. The attached skeleton was not collected because it would have required tunneling into the hillside. AMNH 5373 is a nearly complete skeleton, with a well-preserved skull. Originally Barnum Brown thought that these specimens differed enough from what had been described as *Lambeosaurus* to warrant the creation of a new genus, *Stephanosaurus*. Further research showed that these specimens were the same animal.

ORNITHISCHIAN

Pachycephalosaurus wyomingensis

AMNH: *1696*
AGE: *Late Cretaceous, about 65 mya*
FORMATION: *Hell Creek*
LOCALITY: *William Winkley Ranch, Powder Hill, north of Ekalala, Carter County, Montana*
COLLECTOR: *William Winkley, W. H. Peck, and Thomas G. Nielson*
DATE: *1940*

Although the remains of herbivorous dinosaurs are relatively common, pachycephalosaur fossils are universally rare. *Pachycephalosaurus* and its relatives were small to medium-sized, bipedal herbivores that inherited immensely thick skulls from their common ancestor. In most dinosaurs the bones covering the top of the braincase are thin—at most only a couple centimeters thick—as in other vertebrates. Of the pachycephalosaurs, *Pachycephalosaurus* has the thickest skull, with a roof that is more than 25 centimeters thick. This thick, dome-shaped skull studded with small horns has led to speculation that these animals butted heads in the same way that modern bighorn sheep do during mating season. Supporting this argument is the ability of back and neck vertebrae of pachycephalosaurs to lock rigidly, dispersing the force of a head-on collision throughout the body. These characteristics provide enticing evidence for head-butting behavior, but as with other fossil behaviors, we must be cautious about inferring too much from 65-million-year-old bones.

The size and shape of the skullcap vary depending on the species of pachycephalosaur. In the small Mongolian *Homalocephale*, the top of the head is nearly flat. The head of *Stygimoloch*, whose remains have been found in western North America, was even more highly domed than that of its close relative *Pachycephalosaurus*, and the bony knobs surrounding the dome were enlarged into small horns.

Pachycephalosaurus has small, laterally compressed teeth with small denticles on their edges. In 1856 Joseph Leidy described *Troödon formosus* on the basis of isolated teeth from the Judith River Formation of Montana. Later, because of the superficial resemblance between these teeth and pachycephalosaur teeth, pachycephalosaur material was referred to *Troödon*. Because no relatively complete skeletons of *Troödon* were known at the time, its theropod affinities were not recognized. Not until the late 1940s, when more complete specimens were collected, did paleontologists realize that troödontids and pachycephalosaurs are not closely related, but are very different kinds of dinosaurs (*see "Saurornithoides,"* page 153).

The Museum acquired its *Pachycephalosaurus* specimen in 1941 through exchange with the Carter County, Montana, Geological Society.

FIGURE 108. Although pachycephalosaur skulls are durable, good specimens are rare. *Pachycephalosaurs* were probably uncommon animals even in their heyday. This fine example is the type specimen of *Pachycephalosaurus* from the Hell Creek Formation.

ORNITHISCHIAN

Psittacosaurus mongoliensis

AMNH: *6253, 6254*
AGE: *Early Cretaceous, about 107 mya*
FORMATION: *Oshih and Ondai Sair*
LOCALITY: *near Tsagan Nor Basin, Mongolia*
COLLECTOR: *Won (Chinese driver), Walter Granger*
DATE: *1922*

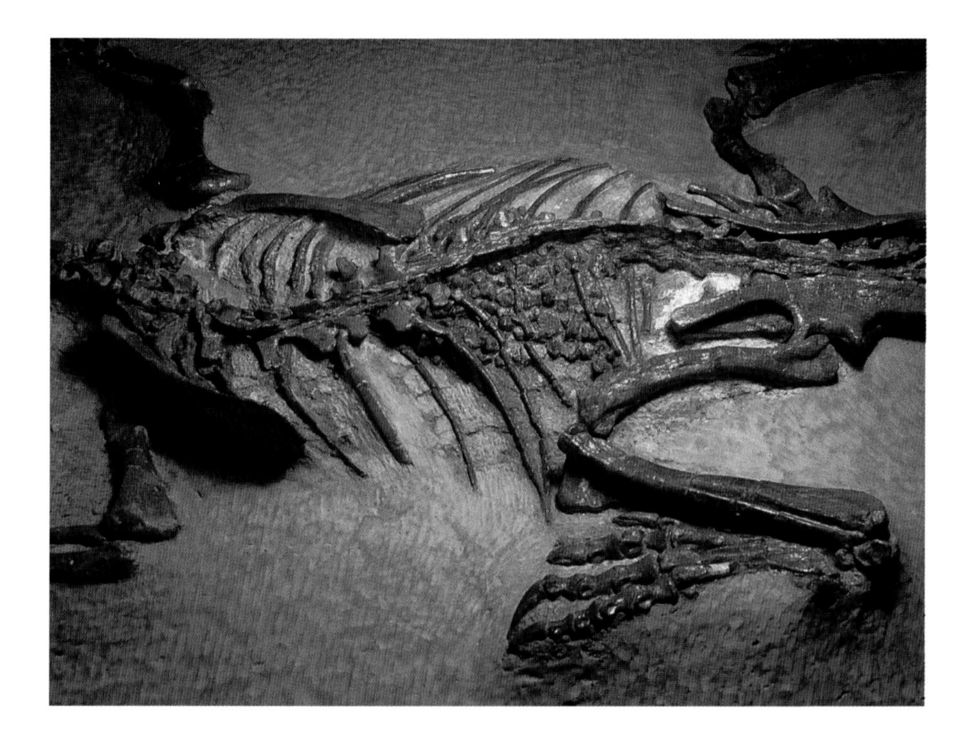

FIGURE 109. Rarely preserved in dinosaurs, the small stones inside the rib cage are probably analogous to the gizzard stones of birds and crocodiles, more correctly termed the gastric mill, a device which presumably aided in digestion.

Psittacosaurus is a relatively small dinosaur, not more than 2 meters long. It is the oldest and most primitive ceratopsian dinosaur. Like other ceratopsians, such as *Triceratops* (*see* page 172) and *Protoceratops* (*see* page 166), *Psittacosaurus* inherited a parrotlike beak, expanded cheeks, and a bony shelf over the back of the skull from their common ancestor. Unlike more-advanced ceratopsian dinosaurs, *Psittacosaurus* was predominantly bipedal and lacked a frill at the back of the skull. The teeth were not developed into batteries as in later ceratopsians, and most of the food processing probably took place in the gastric mill (*see* "What did dinosaurs eat?," page 39). One specimen in the Museum's collection preserves the position of the stone gastroliths inside the rib cage as in life.

In 1922 two remarkably complete and very important skeletons were collected from Early Cretaceous beds in central Mongolia, in the region of the great lake and dune fields of Tsagan Nor. At the time these specimens were thought to represent two different species. Both were described by Osborn in his 1923 paper *"Two lower Cretaceous dinosaurs of Mongolia."* In this paper Osborn designated AMNH 6253 as the type specimen of *Psittacosaurus mongoliensis* and AMNH 6254 as *Protiguanodon mongoliense*. The features that Osborn and other workers used to differentiate these species included minute morphologic details and a perceived *"wide geographic and temporal separation."* The anatomical details have not stood the test of reanalysis. In addition, because many species are known to range over broad geographic areas and to exist for significant periods of time, stratigraphic and geographic separation are not valid criteria for taxonomic distinction. The evidence for how much these two localities differ in age is unconvincing in any case. Even Osborn recognized that the grounds for separating these two species were trivial, cautioning, *"In case the animals turn out to be the same,* Psittacosaurus *will have precedence."*

In addition to these specimens, the 1923 expedition recovered the skulls of two juvenile *Psittacosaurus* at Oshih. When the specimens were found the excavators thought that they had uncovered only a few bone fragments in rock. Preparation in New York revealed the two juvenile skulls and indicated that these investigators unknowingly may have discovered the first dinosaur nest with associated juveniles.

ORNITHISCHIAN

Protoceratops andrewsi

AMNH: *6251, 6417, 6467*
AGE: *Late Cretaceous, about 72 mya*
FORMATION: *Djadoctha*
LOCALITY: *Flaming Cliffs, Mongolia*
COLLECTOR: *Central Asiatic Expeditions*
DATE: *1922*

The first dinosaur discovered at the Flaming Cliffs was a partial skull of the small neoceratopsian dinosaur *Protoceratops* (AMNH 6251). Neoceratopsians, which include *Triceratops, Styracosaurus,* and their relatives, are a large group of Asian and North American herbivores that inherited distinct frills at the back of the skull from their common ancestor. *Protoceratops* is a very primitive member of this group and lacks horns on the nose. Because *Protoceratops* is so primitive relative to other neoceratopsians, it was thought to have occurred earlier in time. Today we know that this belief was mistaken;

FIGURE 110. ***Protoceratops*** **is a primitive horned dinosaur. This animal was made famous through the Central Asiatic Expeditions and has been wrongly identified as the layer of the dinosaur eggs found at the Flaming Cliffs.**

Protoceratops was simply a diminutive, primitive member of the neoceratopsian group that lived in Asia at about the same time as larger, more advanced North American forms.

In the Late Cretaceous fossil beds of southwest Mongolia, *Protoceratops* is often the most common dinosaur skeleton encountered—so common that the Central Asiatic Expedition of the 1920s collected entire skeletons of a wide range of individuals—from large adults to small juveniles. Many of the skeletons were exquisitely preserved; a few are even preserved in death poses, lying on their backs with their feet to the sky.

The extensive growth series of *Protoceratops* specimens make it one of the best-known dinosaurs. As with *Hypacrosaurus* (*see* page 159), the series documents allometric shape changes during growth. This growth series becomes more and more complete as new discoveries are made. A few tiny hatchlings with skulls less than 3 centimeters long were recently collected by the joint expedition of the Mongolian Academy of Sciences and the Museum during the 1994 field season.

In life *Protoceratops* resembled a large pig in size and build. Its teeth indicate that it was an herbivore. *Protoceratops* is probably known best for eggs found at the Flaming Cliffs in 1923, which were originally refered to it. Recent finds definitively show that these eggs do not belong to *Protoceratops* (*see*, "The Flaming Cliffs Dinosaur Nests," page 177, and "The Western Gobi," page 209). At least one specimen (*see* "Did dinosaurs fight?," page 42) suggests that *Protoceratops* was a common prey item of its many carnivorous contemporaries, including *Velociraptor*.

ORNITHISCHIAN

Centrosaurus apertus

AMNH: *5351*
AGE: *Late Cretaceous, about 72 mya*
FORMATION: *Judith River*
LOCALITY: *Red Deer River region, Alberta, Canada*
COLLECTOR: *Barnum Brown and Peter Kaisen*
DATE: *1914*

Advanced neoceratopsian dinosaurs form a group called the Ceratopsidae. Two distinct subgroups within the Ceratopsidae have been recognized: the Centrosaurinae and the Chasmosaurinae. Centrosaurines have a shorter frill and are smaller than their chasmosaurine relatives and have a single large nose horn and very small brow horns above the eyes. There is a great deal of variation in the shape of the nose horn. *Centrosaurus* is thought to be very closely related to *Styracosaurus* (*see* page 170). As we will discuss, the classification of these animals has been the source of a great deal of confusion.

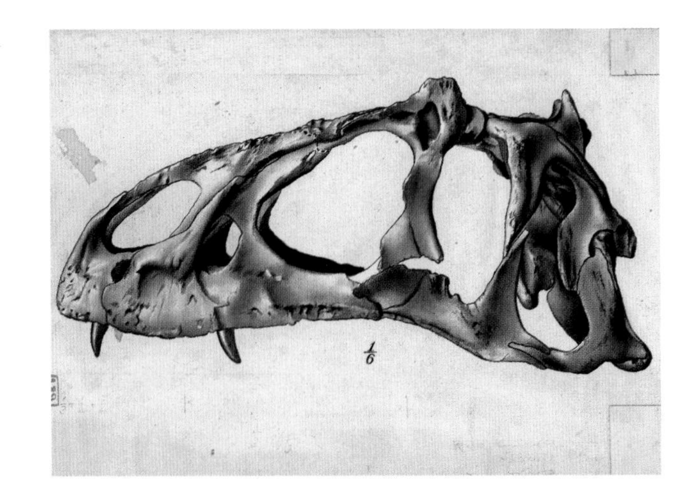

FIGURE 111. This specimen was mounted in a plaque mount in the American Museum of Natural History halls. It is complete except for a few foot bones. Even the bones that support the eye ball (the scleral ossicles) are preserved.

In 1914 Barnum Brown and Peter Kaisen found what Brown considered to be the most complete dinosaur skeleton he ever found. As Barnum Brown explains this specimen of *Centrosaurus* (AMNH 5351) is, *"...complete in all details from the tip of the tail to the end of the nose with most of the bones articulated in position. It was lying on its left side with the phalanxes exposed and some of the bones were damaged but parts of all were present. Part of the skeleton was surrounded by sandstone and ironstone matrix of such nature that much of the scientific nature of the skeleton would have been sacrificed by extracting it for a free mount. Consequently the skeleton has been worked out in relief and mounted as a plaque."*

AMNH 3999, collected on the Judith River by E. D. Cope and Charles Sternberg in 1877, was originally part of the Cope collection. In the 1870s these collection expeditions were long, arduous trips, for the West was fundamentally different from what it would be even 20 years later when the American Museum of Natural History's own field parties began to collect fossils in the area. First, there were few lines of supply; the early paleontologists had to rely on intermittent contact with steamships traveling through the Missouri River

FIGURE 112. One of the finest dinosaur specimens in the Museum's collection is the *Centrosaurus* specimen discovered by Barnum Brown on the Red Deer River. This 1914 tinted plate shows Brown shortly after its discovery. Collection of this specimen was a monumental undertaking. A huge hole needed to be excavated to ensure that all the skeletal remains were recovered.

basin, or on poorly equipped army posts on the prairie. Second, the ongoing Indian wars raging in central Montana during these years posed an immediate danger. The Custer battle took place in 1876, and the Judith River country was on the Sioux escape route to Canada. Consequently, tensions in the region were at a near-panic level. Fortunately, most of the Native Americans encountered during Cope and Sternberg's western foray were friendly bands of Crow and their allies. Occasionally the excavators encountered bands of Sioux warriors, but without serious incident.

On display at the Museum is another skeleton, collected on the Red Deer River, of a medium-sized ceratopsian that may be referable to *Centrosaurus*. What is remarkable about this headless skeleton is that it is preserved as a mummy. Skin imprints of horned dinosaurs are extremely rare, and no ceratopsian specimen as complete as the *Edmontosaurus* mummy is known. This thin skin imprint of what may be *Centrosaurus* shows that ceratopsians were covered with scales in the form of large round tubercles surrounded by smaller polygonal tubercles. The large tubercles seem to have been arranged in rows. The skin imprint is probably from the underside of the animal.

Unfortunately, we cannot tell for sure if this animal is *Centrosaurus*, because all the medium-sized ceratopsians are extremely similar. To identify the animal we need a skull associated with the skeleton, because the ornamentation of the horns and frills constitute most of the diagnostic characteristics of these species. Sexual dimorphism, however, may make even the identifications based on skulls questionable (*see "Styracosaurus,"* page 170).

ORNITHISCHIAN

Styracosaurus albertensis

AMNH: *5372*
AGE: *Late Cretaceous, about 72 mya*
FORMATION: *Judith River*
LOCALITY: *Red Deer River region, Alberta, Canada*
COLLECTOR: *Barnum Brown and Albert Johnson*
DATE: *1915*

One of the more spectacular of the horned dinosaurs is *Styracosaurus*, a dinosaur about the same size as *Centrosaurus* but having, in addition to the large nose horn, a series of enlarged spikes around the edge of the frill. The horns above the eyes, however, are merely small nubbins. Skeletons of *Styracosaurus* are relatively rare; the Museum has the only mounted skeleton. Although nearly all of the postcranial skeleton was found, most of the skull in this specimen was missing or heavily damaged. The skull on the museum's mount is based on a specimen at the Canadian National Museum in Ottawa.

The difficulty in distinguishing ceratopsian species has been exacerbated by paleontologists' enchantment with the shape of the horn and frills. These difficulties were obvious to Barnum Brown who cautioned, *"Most writers in describing species of* Ceratopsia *have attached greater importance to the development of horns and accessory frill growths than to me seems warranted. These parts are subject to great individual variation and too much stress should not be laid on such characters."*

Nonetheless, most of the characteristics that separate *Styracosaurus* from other species have to do with horn shape and size. No living birds or other reptiles display such a variety of horn shapes as ceratopsian dinosaurs.

FIGURE 113. *Styracosaurus* is a short-frilled ceratopsian with spikes around the frill, no brow horns, and a single nose horn. The frill spikes make *Styracosaurus* distinctive; no other ceratopsian has them. *Styracosaurus* was known originally from the Cretaceous sediments in the Red Deer River area of Alberta, but new species have been found in Montana. This specimen is shown as it was being cast in the 1920s.

Paleontologists must therefore turn to mammals to examine putative analogues. In mammals with bony head ornaments like deer and rhinoceros, pronounced differences in horn shape exist between males and females. Males are also usually larger than females and have bigger, more elaborate horns. Such morphologic differences between the sexes are called sexual dimorphism.

Observation of sexual dimorphism in living mammals has led some paleontologists to argue that many ceratopsian species may just be males and females of the same species. This possibility is especially likely when the two species occur in the same deposit. *Styracosaurus albertensis* and *Monoclonius dawsoni* are two of the animals at the heart of this controversy. Citing similarities in the skeleton, some paleontologists have suggested that these are a single dimorphic species. A definitive test of this idea, however, would require statistical analysis involving more fossil specimens of many different size classes than are currently known.

ORNITHISCHIAN

Triceratops horridus

AMNH: *5039 (lower jaws); 5045 (neck vertebrae); 5116 (skull); 970, 971, 973, 5033 (remainder of skeleton)*
AGE: *Late Cretaceous, about 65 mya*
FORMATION: *Hell Creek and Lance*
LOCALITY: *Northeastern Montana and Wyoming*
COLLECTOR: *C. H. Sternberg, Peter Kaisen, and Barnum Brown*
DATE: *1909*

T*riceratops* is one of the best-known dinosaurs. Its remains are so common that Barnum Brown once remarked: *"During 7 years' work, 1902–1909, in the Hell Creek Beds of Montana, I identified no less than five hundred fragmentary skulls and innumerable bones referable to this genus."*

Triceratops and its chasmosaurine relative *Chasmosaurus* inherited a very large frill from their common ancestor. The importance of the characteristic frill and horns of ceratopsian dinosaurs has been something of an enigma to paleontologists. The obvious function of the horns and head shield, and the one usually advocated, is as a defensive structure, similar to the horn of a rhinoceros. This view is supported by the observation that many specimens show patterns of damage to the horns and frill that healed during the animal's life. Probably these wounds were encountered during fights with rivals or battles with predators. In addition, many of the neck vertebrae are fused into a single "cervical bar," ostensibly for stabilizing the neck during frontal ramming attacks. Other explanations are possible, however. For example, the cervical bar could be simply a modification compensating for the immense size and weight of the head.

The extreme variation in the shape of the frill among different, closely related forms has led to other ideas. One early suggestion was that the frill formed an attachment point for large jaw muscles, presumably resulting in a stronger chewing mechanism

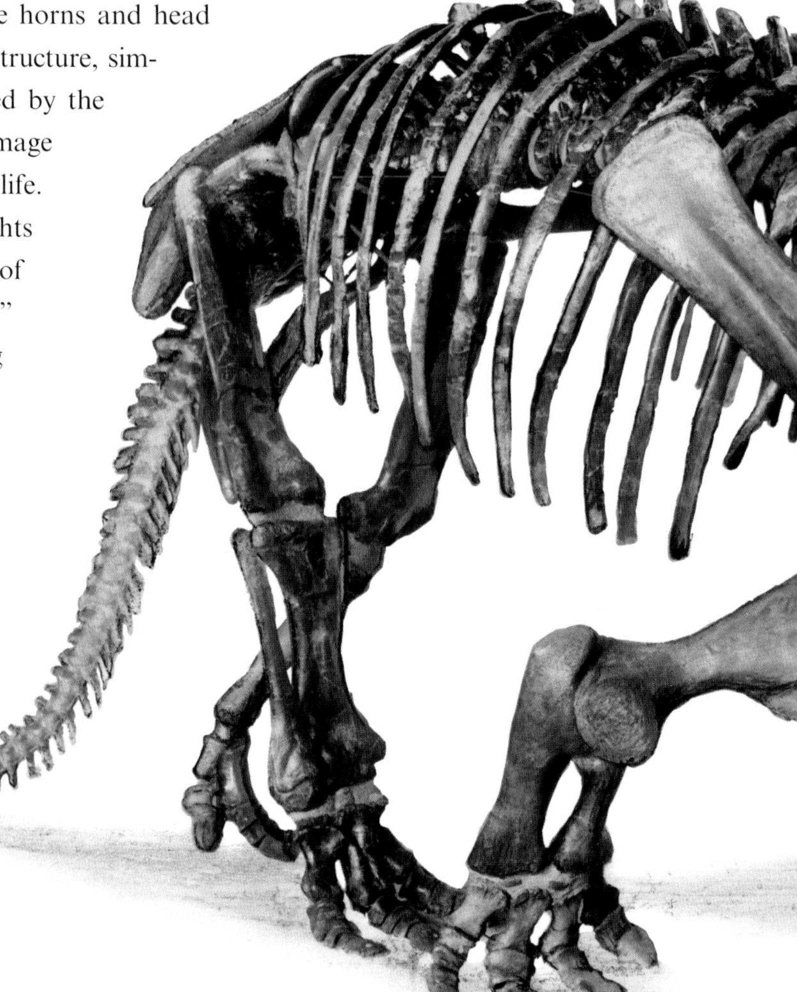

Later it was proposed that the frill was a mechanism for heat dissipation, or that, in association with the horns, the frill was involved in threat or display behavior. This last idea has precipitated popular reconstructions of these animals with fanciful color schemes on the frill and horns, reminiscent of the coloration of moths. In extant animals such coloration, when ascribed to threat behavior, usually involves mimicry. It is difficult to imagine what a 9-meter-tall animal would be mimicking.

The posture of *Triceratops* is also controversial. Traditionally, *Triceratops* and its relatives have been portrayed as slow-moving sprawlers with splayed-out front legs. Many contemporary reconstructions, however, show these animals as upright runners with straight legs held right under the body. Published studies of ceratopsian biomechanics support both possibilities; the only direct evidence, however, comes from ceratopsian trackways.

Footprints are physical evidence for the gait and posture of ceratopsians while alive. By measuring the length of stride and the distance between right and left sides, investigators can determine whether ceratopsians sprawled or walked upright, as well as how fast they moved in relative terms. Ceratopsian trackways that have already been discovered support the theory of an upright posture but do not suggest fast running.

FIGURE 114. Traditionally *Triceratops* and other ceratopsians have been portrayed with their forelimbs splayed out, as in this 1923 mount. Recent reinterpretations suggest that ceratopsians were rapid runners with erect limbs. Unfortunately, the few serious studies of ceratopsian locomotion suggest a range of motion that permits either of these theories, leaving the issue ambiguous.

ORNITHISCHIAN

Chasmosaurus kaiseni/ Chasmosaurus belli

AMNH: *5401, 5402*
AGE: *Late Cretaceous, about 72 mya*
FORMATION: *Judith River*
LOCALITY: *Judith River, Red Deer River region, Alberta, Canada*
COLLECTOR: *Barnum Brown and Peter Kaisen*
DATE: *1912, 1913*

Unlike *Triceratops*, *Chasmosaurus* has an extremely long frill with large openings. Although the nasal horns in *Chasmosaurus* are small, the horns above the eyes range from extremely long to virtually absent. *Chasmosaurus* is the smallest of the chasmosaurine ceratopsians. Its fossil record is extraordinary; bone beds that contain numerous individuals (sometimes more than a hundred) of the same species are common. These occurrences have been the root of speculation regarding possible gregarious behavior in these animals, similar to that of large herbivores today. The mass mortality is described as the result of a natural calamity such as fording a swollen river during migration.

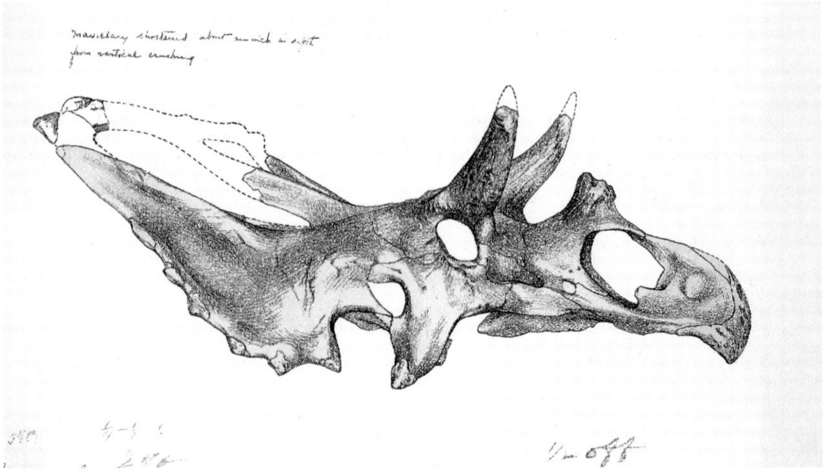

A recent statistical analysis of *Chasmosaurus* specimens suggests two morphologically distinct groups can be distinguished on the basis of horn size. These may be males and females of the same species, indicating that *Chasmosaurus* was sexually dimorphic. Detailed statistical analyses of these samples are only beginning, but they promise to reveal important details about ceratopsians (such as the amount of variation in a species and how animals changed allometrically during growth). In turn, these studies may clarify some classification problems within other ceratopsian groups, such as whether *Styracosaurus* and *Centrosaurus* are one type of dinosaur or two. Like some of the hadrosaurs and armored dinosaurs, *Chasmosaurus* thrived along the margins of the Cretaceous epicontinental seaway (*see* "Why did nonavian dinosaurs become extinct?," page 62).

FIGURE 115. *Chasmosaurus* is one of the long-frilled ceratopsians. Like *Pentaceratops*, its frill extends backward as a light, bony framework, with two large openings. The brow horns of *Chasmosaurus* were large and its nose horn small or absent. As with so many dinosaur characteristics, the function of the frill in ceratopsians is unknown. Current thinking favors sexual display and recognition of species, but these theories are mere speculation.

TRACKS AND EGGS

The Connecticut Valley Tracks

AMNH: *28490*
AGE: *Late Triassic, about 220 mya*
FORMATION: *Newark Supergroup*
LOCALITY: *Turners Falls, Massachusetts*
COLLECTOR: *James Deane*
DATE: *1845*

In 1802 Pliny Moody, a Massachusetts farm boy, plowed up pieces of flagstone covered with the impressions of fossil footprints in his family's fields near Hadley, Massachusetts. These trackways became locally known as the footprints of "Noah's Raven." Such footprints are common in New England, yet they escaped scientific attention for more than 30 years after Moody's discovery, until James Deane, a doctor and naturalist, noticed fossil tracks resembling turkey footprints in rock slabs destined for roadbeds near his home. In 1835 he wrote Reverend Edward Hitchcock of Amherst College of his discovery, *"In slabs of Sandstone from the Connecticut River, I have obtained singular appearances, new to me, although I presume not to yourself. One of them is distinctly marked with the tracks of a turkey (as I believe) in relief. There were two birds side by side making strides of about two feet.... The tracks, four in number, are pefect, and must have been made when the materials were in a plastic state, and at what period I leave you to tell."*

Hitchcock secured these for his collection at Amherst College. He went on to make a career of describing the tracks and published his results in lavishly illustrated monographs. In his initial paper, which appeared in 1836, Hitchcock assigned tracks to all sorts of animals, known and unknown. Some of the footprints Hitchcock believed were made by mammals related to living marsupials such as opossums. Because most of the tracks were thought to be those of large birds, however, they became known as "Hitchcock's ravens."

FIGURE 116. **Footprint slabs discovered by the early collectors, such as this one from the Museum's collection, found their way into most of the old northeastern U.S. natural history collections. This specimen, collected near Turner's Falls, Massachusetts, by J. Dean in 1845, shows the footprints of several kinds of small dinosaurs.**

FIGURE 117. Fossil dinosaur footprints from western Massachusetts were widely known as local curiosities in the early 1800s, as this Currier and Ives lithograph attests.

As the dinosaur tracks of the Connecticut Valley became better known, Deane and Hitchcock both tried to take credit for the discovery. Claims and counterclaims became so heated that much of Hitchcock's later career was resigned to defending himself, and shelling out vitriolic counterattacks against charges of impropriety made by Deane. Even after Deane's death in 1858, Hitchcock continued to pen lengthy dispositions concerning his right to the claim of first scientific descriptor of the tracks.

Although Deane and Hitchcock were wrong about what kinds of animals made the Connecticut Valley tracks (not until the mid-1860s, after many European discoveries, did general consensus acknowledge them to be the footprints of dinosaurs), both were careful investigators. They described features as subtle as raindrop impressions. Hitchcock and Deane also possessed enlightened views about the relationship between their spiritual beliefs and scientific findings. Statements such as Hitchcock's 1863 remark that *"the real question is, not whether these hypotheses accord with our religious views, but whether they are true"* are still considered by many to be heretical.

Remarkably, these scientists used comparative methods similar to those of good modern paleontologists. Consider this passage from Deane's 1861 posthumously published monograph: *"If living animals be found whose footprints conform in every essential particular to the fossil impression, it must in reason be conceded, that the organization and habits of the extinct and living types are also conformable."* Unknowingly, Deane and Hitchcock were the first dinosaur trackers to examine dinosaur paleobiology empirically. Hitchcock measured stride lengths to estimate relative speed in different animals, examined gaits of his various "orders," and even speculated, on the basis of the orientation of groups of tracks, on the social habits of the track makers.

Fossil footprints are still commonly found throughout New England. The much rarer occurrence of dinosaur bones makes it difficult to tie fossil tracks to specific kinds of dinosaurs. The first dinosaur bones collected in the Connecticut Valley (a partial skeleton of the prosauropod *Anchisaurus*) were found in East Windsor, Connecticut, in 1818. Only a handful of dinosaur skeletons have been recovered from these strata since then.

The Flaming Cliffs Dinosaur Nests

AMNH: *6508, 6509, 6631*
AGE: *Late Cretaceous, about 72 mya*
FORMATION: *Djadoctha*
LOCALITY: *Flaming Cliffs, Mongolia*
COLLECTOR: *Central Asiatic Expeditions*
DATE: *1923*

In 1923, to worldwide fanfare, the Museum announced the discovery of the first bona fide dinosaur eggs. The first egg had been found at the Flaming Cliffs in Mongolia by Central Asiatic Expedition photographer J. B. Schackelford in September 1922. At first, it was thought to be a fossilized bird egg. Because the abundance of dinosaur fossils at the locality indicated the deposit to be Cretaceous in age, however, it was not long before the excavators began to speculate on the possibility this was the egg of a nonavian dinosaur.

When full-scale expeditions returned to the Flaming Cliffs in 1923 for a summer of excavation, the collectors discovered entire nests of eggs. More were found in 1925. During the 1925 expeditions, Roy Chapman Andrews, never one to miss a publicity opportunity, shot extensive cinema footage of the excavation. In one of these "shorts", Andrews and Granger (the head paleontologist) restaged the discovery of the nest of dinosaur eggs two years earlier. The grainy black-and-white film makes these eminent explorers look more like keystone cop rejects than the discoverers of one of the great fossil localities of all time.

The discovery of these eggs generated copious popular interest, which Andrews used in procuring funds for future Asian expeditions. Among his promotional techniques were special viewing parties for New York socialites and potential contributors. Popular articles soon followed, and casts and specimens were exchanged with other museums worldwide.

In one regrettable publicity stunt, an egg was auctioned to the highest bidder. Only wealthy benefactors were allowed to bid, with the stipulation that the specimen go into the collection of a major university or museum. Inquiries came from all over. The highest bidder was Colonel Austin Colgate, who purchased the egg for $5,000 and presented it to Colgate University, where it resides today. What made the incident problematic was that the auction generated enough publicity that antagonists of the expedition in the Mongolian government made use of this episode to discredit the Central Asiatic

Expeditions as a for-profit venture intended to steal Mongolia's patrimony.

Most of the eggs collected by the Central Asiatic Expeditions at the Flaming Cliffs are the same oblong shape, about 18 centimeters long, with one end thinner than the other. The surface of the eggs is rough, covered by small tubercular ridges. Many were found in nests. Unlike many egg sites elsewhere in the world, no local aggregations, suggestive of communal nesting sites, are apparent. Several other types of fossilized eggs were collected at the Flaming Cliffs by Museum paleontologists and others, and eggs are still common at the site today.

FIGURE 118. A nest of dinosaur eggs found by the Museum's Central Asiatic Expedition in Late Cretaceous sediments at the Flaming Cliffs locality in Mongolia. The original notes with this image say that this nest is shown as found; in all likelihood, however, these eggs were specially arranged for this photo opportunity.

To what dinosaur do these eggs belong? *Protoceratops* was a likely contender (*see* page 166) for laying the most common type of egg. Besides its abundance at the Flaming Cliffs, however, no evidence suggests that the eggs from the Flaming Cliffs are *Protoceratops* eggs. No *Protoceratops* bones have been found in direct association with the eggs encountered at the deposit, and no embryonic *Protoceratops* (the only definitive evidence) have been found inside the eggs. Instead, some of these nests have been shown to belong to oviraptorid dinosaurs (*see "Oviraptor,"* page 124, and "The Western Gobi," page 209).

These nests tell us more than just that dinosaurs laid eggs. Eggs are arranged in the nests in a definite pattern with their narrow end facing the outside and the broad end facing inward. This is very unlike pattern found in nests of primitive reptiles like sea turtles where the eggs are laid almost randomly in an excavation on a beach. Because it would be very difficult to lay the eggs in such a regular orientation, adult dinosaurs were probably responsible for manipulating the eggs in the nests. Such behavior is found in modern birds. The non-random pattern of eggs in these nests attests to the close relationship between birds and nonavian dinosaurs and points out that many of the behaviors seen in living birds have roots that were inherited from common ancestors very deep in nonavian dinosaur history.

The Coal Mine Tracks

AMNH: *3650*
AGE: *Late Cretaceous, about 75 mya*
FORMATION: *Mesa Verde*
LOCALITY: *United States Mine, Cedaredge, Colorado*
COLLECTOR: *Barnum Brown and R. T. Bird*
DATE: *1937*

FIGURE 119. The footprints in the United States Mine were actually the fillings of footprints preserved on the mine's ceiling. This unusual position made the job of collecting the tracks a monumental struggle.

In 1937 Barnum Brown received reports of the discovery of giant dinosaur footprints in the Chesterfield Mine near Sego, Utah. Brown identified these tracks as belonging to a new, mysterious animal that he called the "Mystery Trackmaker." He immediately began searching for bones in the area, hoping to find remains of this huge animal. Brown was able to conscript the services of the Union Pacific railroad, who donated a steam shovel to the project. The Museum crew, with the aid of the steam shovel, went to one of Brown's prospects and began excavating. Much to their disappointment, they found almost nothing. Soon Brown's attention was diverted by other reports of large trackways found deep within the coal mines of western Colorado.

On visiting the site, near Cedaredge, Colorado, Brown and R. T. Bird were met by Charles States, the mine's owner. After a long journey into the mine, they found footprints of the "Mystery Trackmaker" exposed on the ceiling. States provided manpower and equipment for the dangerous job of collecting the tracks. Rather than depressions, the tracks were positives, formed when sand filled footprints that had been made in a mat of dense vegetation. The vegetation eventually turned into lignite (a form of coal) and was removed during mining operations, leaving positive impressions of the tracks exposed on the ceiling. What was exposed was not a footprint, but rather the petrified filling of the track itself.

The thick sandstone in which the trackway was preserved had to be thinned, a time-consuming process, before it could be removed and transported. Finally, each of the tracks was removed individually. At the American Museum of Natural History the slab was reconstructed on a vertical wall, where it still stands outside the Hall of Ornithischian Dinosaurs.

Brown considered this animal to be the largest dinosaur of them all, a lumbering brute approaching 10 meters in height. He based his claim on the stride of the animal, which he measured at more than 5 meters. In subsequent years, after the tracks had been identified as belonging to a hadrosaur, the long stride length was cited as conjectural evidence that these animals were moving very quickly. In 1981 Tony Thulborn, a specialist in dinosaur locomotion, noted that Brown had missed a track between the two that he had measured. The actual stride length was a mere 2.5 meters, one-half of Brown's original estimate. Consequently, the United States trackway represents poor evidence for speedy dinosaurs (*see* "How fast did dinosaurs move?," page 33).

Brown and Bird made several other discoveries during their work in the Mesa Verde coal mines. Many of the local old-timers recalled occurrences of beautiful imprints of fossil plants in the Red Mountain Mine, adjacent to the United States Mine. One particular gallery, known as the picture room, contained imprints of large palm fronds. The Red Mountain Mine had been closed for years and had caved in substantially. At considerable personal risk, Bird convinced the mine foreman and Charles States to take him into the decaying mine shaft to locate these leaves.

Most of the specimens either had been destroyed through attrition or existed in a more pristine state in the memories of the miners than in reality. One spectacular specimen, however, was located near the bottom of the tunnel. After extensive retimbering and several close calls, R. T. Bird succeeded in acquiring the palm leaf, (*see* "What was the world like during the time of dinosaurs?" page 19). R. T. Bird describes the scene: *"Randomly, at varied and completely unpredictable intervals, lumps of rock had dropped from ceiling to floor all during the operation, as lumps of rock have ever done in old mines. Rock falling on rock is a matter of no moment in the out-of-doors. But in the underground, if it happens well away, this creates a spooky thunder that rolls and rolls.... Three hours later we moved into the area where the dinosaurs had made a playground out of our roof. The long room with the badly splintered props materialized out of the velvet gloom. We laid our planks with greatest care across the dusty floor. Inch by inch we edged and eased along; we didn't touch a thing on either side. We shivered when the rollers squeaked or squeal. The footprint-pitted ceiling floated overhead."*

TRACKS AND EGGS

The Paluxy River Trackway

AMNH: *3065*

AGE: *Early Cretaceous, about 107 mya*

LOCALITY: *Glen Rose Formation, Texas (Paluxy River, Glen Rose)*

COLLECTOR: *R.T. Bird*

DATE: *1938*

Throughout the 1930s and early 1940s, R.T. Bird roamed the western United States on his Harley-Davidson, searching for fossils for the American Museum of Natural History. Bird had come to the Museum to be Barnum Brown's apprentice, working in the preparation laboratory and assisting Brown in the field. During this time, however, Brown was overcommitted as he tried to raise funds for dinosaur collecting by doing contract work for oil and mining companies, so Bird was alone on many of his prospecting forays.

In 1937 Bird was searching the badlands of northern Arizona and New Mexico, exploring possible leads for Brown, when he was informed of the presence of fossil footprints that had been taken to an Indian trading post in Gallup, New Mexico. Upon examining the unusual prints, which seemed almost human, Bird decided they were forgeries. The owner of the tracks tipped him off to another set of footprints, however, which turned out to be definitely authentic. Bird learned that both of the specimens came from the Paluxy River area in central Texas. Immediately he made plans to visit the region.

When Bird arrived in Glen Rose, Texas, near the Paluxy River, he found that the footprints were well known to local residents. A large theropod footprint was even incorporated into the base of the town bandstand. Bird began talking to local people about the occurrence of tracks on surrounding ranch land. He found not only that tracks were common, but that locals had been quarrying them for years to sell to people building rock gardens. In this highly conservative part of the country, Bird found the ranch owners quite suspicious of his activities and not always sympathetic toward a Yankee paleontologist collecting fossils on their land. Bird's ability to convince people to allow him access to their property is testimony to his patience and perseverance.

In his forays around the Paluxy, Bird made a number of remarkable discoveries. These include the Davenport Ranch Trackway, the controversial

Mayan Ranch trackway, and the Paluxy River Trackways. Because of their huge size and the hardness of the limestone strata in which they are embedded, most of these trackways have never been collected.

At the Davenport Ranch Bird found a sauropod stomping ground that captured an entire herd of various sized animals traveling together. Bird's interpretation of this site made him the first to suggest that sauropods moved in structured herds (*see* "Did dinosaurs travel in herds?," page 45).

The Mayan Ranch Trackway is a set of prints that appear to represent only the front feet of a large sauropod (*see* "Were dinosaurs aquatic?" page 57)). Bird interpreted this trackway as that of a swimming sauropod, buoyed up by the water, using only its front feet to push itself along the bottom. Bird's interpretation is tied to the then prevailing idea, that sauropods were aquatic animals that required water to support their immense bodies. The idea that these animals lived in water was also used to explain the lack of tail drags behind sauropod footprints, because the tails would have been floating behind the animals.

Contemporary analysis of the Mayan Ranch Trackway supports an alternative theory—namely, that this trackway is simply poorly preserved. In most trackways of quadrupedal animals the forefeet leave a deeper imprint than the hind feet because the front half of an animal weighs more than the hindquarters. The diameter of the front feet is also less, so the weight is more narrowly dispersed, and the pressure per square centimeter is higher than on the hind feet. These factors combine to make a deeper front footprint in a loose substrate. If, as in the case of the Mayan Ranch Trackway, an animal makes a trackway that is subsequently subjected to weathering and erosion,

FIGURE 120. The main part of the trackway that was excavated is shown here and depicts a large brachiosaur followed by a theropod. This is only one segment of several parallel trackways that may indicate a herd of brachiosaurs being pursued by several theropods. The immense size of this limestone slab, combined with its position in the bottom of a flowing river, made its excavation very difficult. The sand bags that formed a coffer dam diverting the Paluxy River are visible at the lower left.

a deeper level is preserved where only imprints of the deepest footprints (called underprints) remain. In Martin Lockley's reanalysis of the Mayan Ranch Trackway, this appears to be the case. In fact, faint impressions of hind feet, which Bird missed in his original description of the trackway, are visible. No fossil evidence supports the idea that there were swimming sauropods, although lack of evidence doesn't necessarily mean that dinosaurs could not swim to some degree.

Bird's most famous discovery is a theropod trackway on the banks of the Paluxy River. The trackway led into the river, becoming submerged under shallow, running water. To collect the specimen, the river had to be diverted away from the trackway. The hard rock and hot temperatures contributed to making this a monumental job. Bird was able to contract a local Works Progress Administration crew to perform the arduous labor. During the process, the footprints of a large sauropod were found in conjunction with those of a theropod. By the time this excavation was completed, more than 7 tons of rock had been shipped east, yet only a small segment of the trackway had been collected.

The Paluxy Trackway is composed of the tracks of at least twelve sauropods, followed sometime later (as can be determined by the pattern of overlap) by at least three carnosaurs. Based on the spatial relationship between the theropod and sauropod tracks, Bird interpreted the Paluxy slab to be a hunt in progress. Recent analysis suggests that Bird's interpretation may not necessarily be correct. It is impossible to tell exactly how much time passed before the theropods followed; the tracks could represent a hunt in progress, but the theropods may have followed hours or even days after the sauropods.

After his retirement in the early 1940s, R. T. Bird was commissioned by the American Museum of Natural History to reconstruct the Paluxy River Trackway. Today the enormous slab lies behind the renovated *Apatosaurus* mount in the Museum's Hall of Saurischian Dinosaurs. Although dramatic, this mount is in some ways deceiving. The tracks are about 30 million years younger than the dinosaur that seems to walk out of them. Recent study of the trackway has also indicated that *Apatosaurus* is probably the wrong kind of dinosaur to make the tracks. Instead of a diplodocid, the trackmaker was probably a brachiosaur. The theropod footprints, probably do correspond to an *Allosaurus*-like carnosaur. Unfortunately, remains of fossil bones that may bear on this question are extremely rare in the track-producing beds.

Eggs of ?*Hypselosaurus priscus*

AMNH: *7649*
AGE: *Late Cretaceous, about 68 mya*
FORMATION: *Gräs Ö Reptiles*
LOCALITY: *Rousset (Bouche-du-Rhîne), France*
COLLECTOR: *Museum Histoire Naturelle, Aix-en-Provence*
DATE: *1956*

Although dinosaur eggs are becoming increasingly common, eggs that can be attributed to a particular type of dinosaur are still exceedingly rare. The identity of even well-known eggs, such as many of those from Mongolia (*see* "The Flaming Cliffs Dinosaur Nests" page 177) is still up in the air. Consequently, referral of eggs to a particular dinosaur needs to be done with caution. In the case of *Hypselosaurus*, a small Late Cretaceous sauropod found in France, the antecedent "?" means that our referral of the eggs to this dinosaur is based on inference rather than on empirical evidence.

FIGURE 121. Sauropods are the biggest dinosaurs, and it is widely assumed that they laid the biggest eggs. Bones of *Hypselosaurus*, a poorly known titanosaurid sauropod from Cretaceous rocks in France, have been found associated with relatively large, oval eggs. Eggs of this type have been described from Asia and North America as well, but they have not been positively identified as belonging to sauropods.

The only way to determine which dinosaur laid a particular type of egg is to find an identifiable embryo fossilized inside the shell. Only a few dinosaur eggs with embryos have been collected. Unfortunately, none of these are sauropods. Evidence that eggs from southern France are sauropod eggs, and specifically *Hypselosaurus* eggs, is inferential. These inferences are based on the association of the eggs with *Hypselosaurus* bones and on their large size (up to about 25 centimeters in diameter). The argument is that because these eggs are very large, they should belong to members of the dinosaur group with the largest adult body size, the sauropods. Reference to *Hypselosaurus* is also consistent with the fact that *Hypselosaurus* is the largest dinosaur found at any of the egg-containing localities. Is this evidence sufficient for identification? Probably not. Consider the 70-year-long misidentification of the Museum's *Protoceratops* eggs in light of incontrovertible new evidence (*see* "The Western Gobi," page 197). It is better to consider the eggs discovered in southern France as ?*Hypselosaurus*.

The Museum acquired its display specimen of the ?*Hypselosaurus* eggs by exchange in 1957, a year after they were collected

III NEW DISCOVERIES

Confuciusornis sanctus

NGMC *(National Geological Museum of China): various numbers*
AGE: *late Jurassic or Early Cretaceous*
FORMATION: *Chaomidianzi Formation*
LOCALITY: *Sihetun area, P. R. China*
COLLECTOR: *Representatives of the Geological Museum of China*
DATE: *various dates in the 1990's*

Paleontologists rarely find animal fossils that can compare with the specimens of modern animals, either in quality or number. The perverse nature of the fossil record usually results in fossilization of only bits and fragments of interesting animals. *Confuciusornis* is an exception. In the fine-grained sediments of China's Liaoning-Sihetun area hundreds, and perhaps thousands, of these primitive birds have been preserved.

Like a number of the fossils from Liaoning (see "Liaoning," page 214), many of the *Confuciusornis* specimens preserve soft tissue, usually feathers and bits of beak and claw. Often these birds are found as groups, lying adjacent to one another fossilized in rock. This has led to speculation that the animals were killed en mass by the volcanic dust that entombed them.

FIGURE 123 Thousands of beautifully preserved specimens of the Mesozoic avialan *Confuciusornis sanctus* have been discovered in the last ten years.

FIGURE 124 (OPPOSITE) Some specimens of *Confuciusornis sanctus* even preserve tail feathers. On some individuals the tail feathers are extremely long, perhaps suggesting different plumation patterns seasonally or between the sexes.

Occasionally *Confuciusornis* specimens are found sporting a pair of extremely long tail feathers. These tail feathers can be longer than the body itself, like those of some modern-day Birds of Paradise. Because the bone structure between the long tailed and short tailed form is identical, the favored explanation is this plumage difference indicates sexual dimorphism (differences between the sexes). A common phenomenon in modern birds, the more elaborate plumage pattern is usually found in males. If this is true than it can help explain why the long tailed form is rare. In today's birds extreme male plumage pattern usually only appears with full maturity, and is often present only seasonally.

It is easy to get seduced by the beauty of the specimens themselves and be distracted from the profound importance that *Confuciusornis* has in deciphering the interrelationships among early birds. Its anatomy is very primitive, displaying several features found in the even more primitive theropod dinosaurs, but not in more advanced birds. For instance the postorbital bone (the bony bar that defines the back of the eye socket) is complete instead of reduced to a small spike, as in modern birds. A lack of a postor-

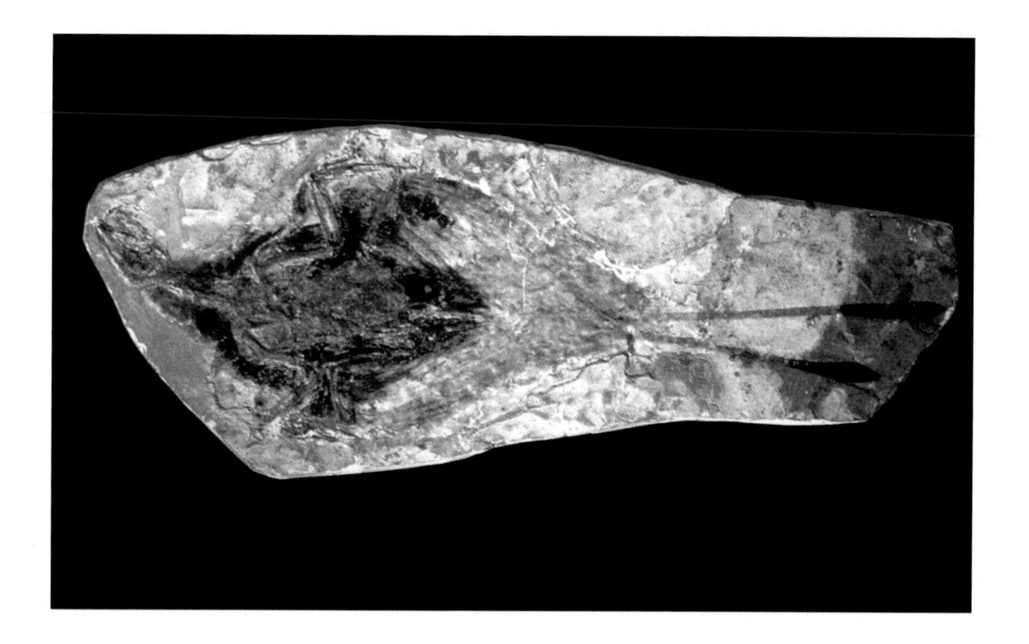

bital bone in modern birds allows a condition called cranial kinesis (which roughly translates to 'skull movement') to occur. This skull movement allows modern birds to raise and lower their beak, by bending some of the skull bones, which has been implicated in advanced feeding strategies. Because the skull of *Confuciusornis* is firmly braced by the bone strut behind the eye (the postorbital), it is impossible that bird-like cranial kinesis was present in *Confuciusornis*.

Confuciusornis is also important in that it is the most primitive bird to show so many advanced features. For instance, the skull lacks teeth and the tail is reduced from the long bony "reptilian tail" seen in *Archaeopteryx*, to a short tail—the pygostyle of modern birds—composed of only a few bones. The reduction of the tail undoubtedly gave *Confuciusornis* a different, more advanced sort of gait than Archaeopteryx. Reduction in the size of the tail requires that the animal's center of gravity shift backward. Consequently in *Confuciusornis* the legs are placed in a more forward position on the body—as in modern birds—otherwise the front of the body would crash to the ground without the long reptilian tail to counterbalance it.

Caudipteryx zoui

NGMC: *(National Geological Musuem of China): 97-4-A*
AGE: *Late Jurassic or Early Cretaceous*
FORMATION: *Chaomidianzi Formation*
LOCALITY: *Sihetun Area, P.R. China*
COLLECTOR: *Representatives of the Geological Museum of China*
DATE: *1997*

When paleontologists first looked at this animal they had no idea what to make of it. It is a weird beast. Some wanted to make *Caudipteryx* a bird, others a close relative of birds, and others a relative of oviraptorid dinosaurs. Many of the reasons for disagreement were superficial. The uninitiated looked purely at the soft tissues that were preserved on the specimen (*see* "Liaoning," page 214). The soft tissues are feathers on its body, including its arms and a large plume on its rear.

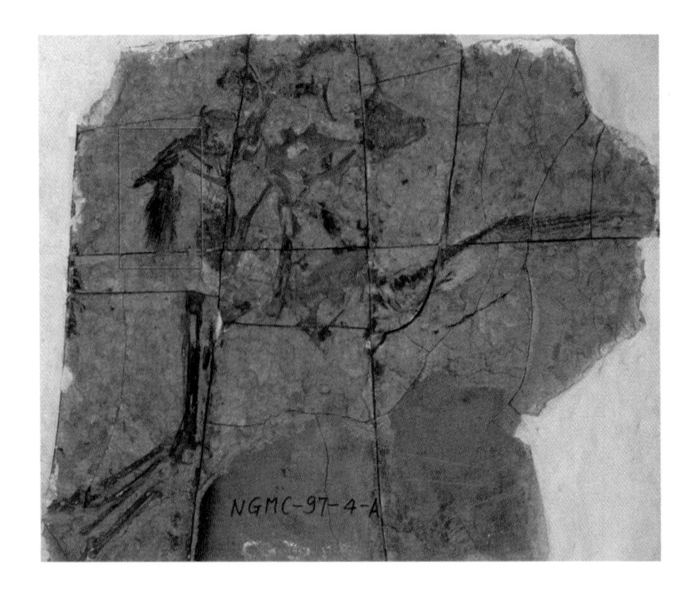

FIGURE 125. Specimens of *Caudipteryx zoui* show long feathers on the tail and vaned feathers on the hand. These "birdlike" attributes are especially surprisng in an animal related to oviraptorid dinosaurs.

Usually we associate feathers with animals that fly. These animals, however, were clearly not flyers. Even though they carried feathers, their forelimbs were too short relative to their body mass to have propelled them through the air. Also the feathers themselves, specifically those distributed across the back of the arms, lack the specializations required for flight in birds. In birds the flight feathers are asymmetric. This means that their leading edge is shorter than the trailing one—a configuration that is mimicked by the wings of airplanes. This configuration forms an airfoil and provides lift. Even the most primitive bird known, *Archaeopteryx*, has airfoil flight feathers. Most paleontologists who have analyzed the matter postulate that *Archaeopteryx* had limited flight capability.

If *Caudipteryx*'s feathers were not for flight, then why do they exist? Other reasons are plausible. One good candidate is that feathers were used for insulation. In modern birds, all of which are warm-blooded, feathers are an insulating covering that provides a thermal barrier between the animal and the environment; they keep even penguins warm against Antarctic tem-

FIGURE 126. The feathers on the forelimb of *Caudipteryx zoui* are identical to the feathers in modern birds, except the vanes are arranged symmetrically around the shafts.

peratures. We also use this natural technology when we don down jackets or snuggle under comforters.

Another possibility is display. Many modern birds use feathers to attract mates or intimidate predators. Peacocks, pheasants, roosters, and cockatoos are common examples. *Caudipteryx*'s large tail plume is especially suggestive, as it is hard to imagine another function for it.

Do its feathers prove that *Caudipteryx* was warm-blooded or that they were used for display? The answer is that a structure's function is notoriously hard to determine in animals that have been dead for over 100 million years.

Even so, the presence of feathers on an animal that is distantly related to birds allows us to make profound predictions. Humans and their closest relatives, chimps, have hair. That is because we share a common ancestor that also had a hairy body covering. This allows us to postulate that protohumans like Lucy and Neanderthals had hair even though there is no fossil evidence that preserves this feature. Similarly, the common ancestor of parrots and *Caudipteryx* had feathers, explaining their presence in both. But this hypothesis has consequences in the sense that it predicts that all of the descendants of the common ancestor of birds and *Caudipteryx* were also feathered. This includes the non-avian dinosaurs like *Velociraptor* and other dromaeosaurs.

This prediction has been borne out by recent discovery. A therizinosaur, a member of the group of dinosaurs postulated to be a close relative of oviraptorid and therefore *Caudipteryx*, was recently discovered that also carried feathers. And there is more to come. At this writing Liaoning and other localities around the planet are giving up their secrets.

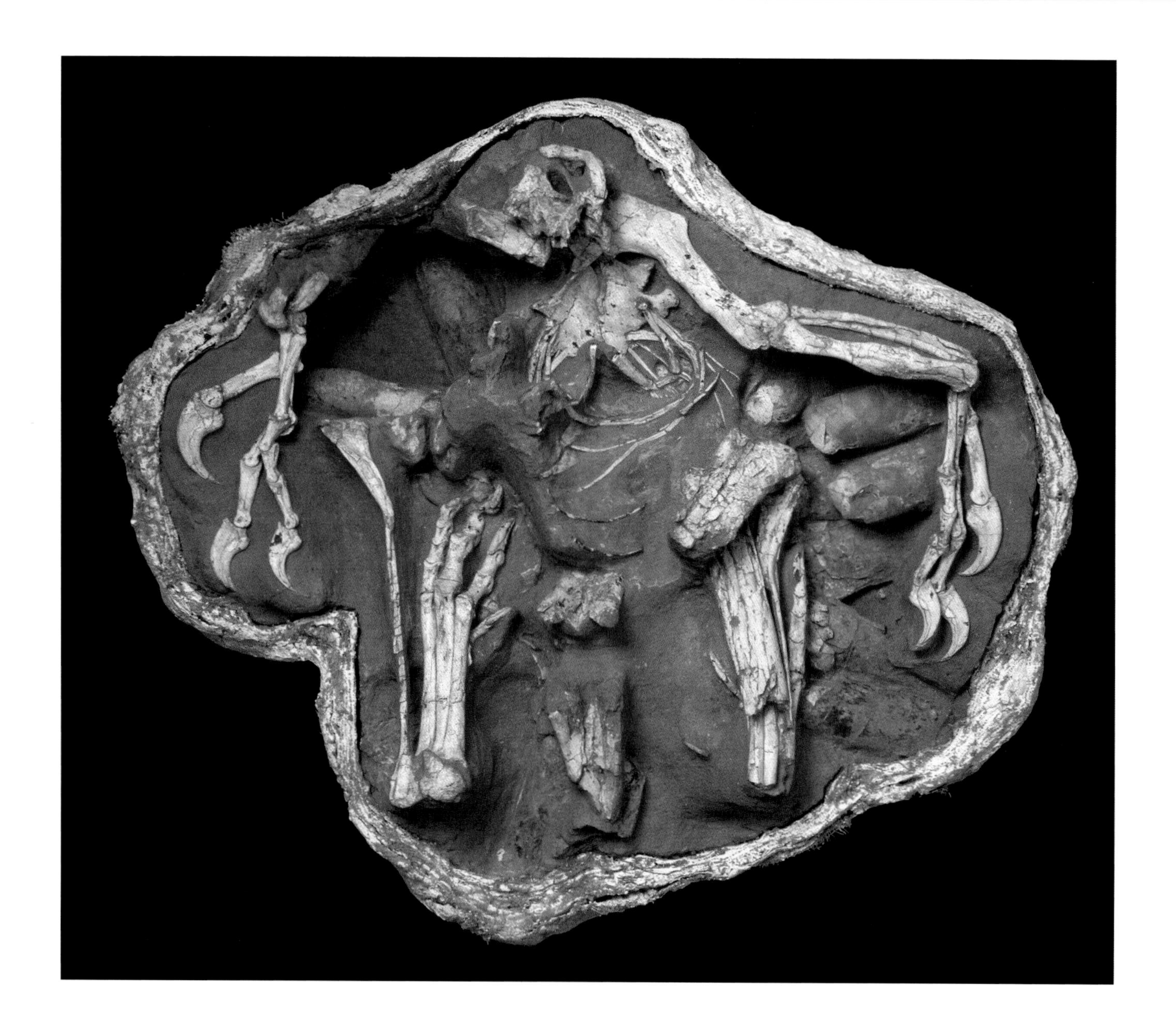

FIGURE 128. The nesting oviraptorid is one of the most important dinosaur specimens discovered in recent history. It preserves an adult on the nest in an identical position to that taken by modern birds while brooding their nests.

oviraptorids near oviraptorid nests are undoubtedly remnants of some behavior. The number of occurrences can not be chance alone. Second, the presence of at least three animals in a brooding position seems to point to a behavior shared among members of these species.

Because the stereotypical behavior of sitting on a nest in this fashion is common to both living birds and these close non-bird relatives, using the methods of cladistics we can predict that this behavior was present in the common ancestor of these animals. Nest-sitting then is something that evolved long before the first birds and was a feature common to several non-dinosaur bird groups—at least all of those that are descended from the last common ancestor of Oviraptor and modern birds. We cannot tell from these specimens whether or not their brooding behavior evolved because these animals were warm blooded (see "*Caudipteryx*," page 188) and the embryos within the eggs required heat for development, or that it originally evolved as a nest-guarding strategy, or even as a way to shade the nest from sun.

NEW DISCOVERIES

Sauropod embryos

PVPH *(Paleontology Collection of the Plaza Huincul Museum, Argentina): various numbers*

AGE: *Late Cretaceous—about 80 mya*

FORMATION: *Río Colorado Formation*

LOCALITY: *Auca Mahuevo, Argentina*

COLLECTOR: *American Museum-Plaza Huincul Museum Field Parties*

DATE: *1997*

Among dinosaurs, the sauropods are probably the most familiar. They are the dinosaurs that embellish gas station signs and whose images form the cover of more kids dinosaur books than any other kind. Sauropods were some of the first dinosaurs to capture the North American public's attention, as it was their remains that fueled much of the great dinosaur bone rush to the American West in the late 19th and early 20th century (*see* "*Apatosaurus*," page 102, and "The Medicine Bow Anticline," page 198).

These animals are also the largest dinosaurs. Because they are so big, they have been the focus of countless studies and debates about dinosaur behavior, growth, and physiology.

Some remarkable claims have been made about them. Early on some scientists thought that these huge animals were so big that they could not possibly have walked on land. Trackway evidence dispelled that conjecture (*see* "Paluxy River Trackway," page 181). More recently it was suggested that some sauropods had up to seven hearts needed to pump blood up their long necks to their tiny brains. Even though no one took that seriously, discussion continues about whether these animals had hearts strong enough to allow them to lift their head above body level.

Why this debate? Although sauropods are known from parts of hundreds, probably thousands of skeletons worldwide, they are still poorly understood. Their large size allows some bones to preserve easily. Relatively few complete skeletons, however, have been found. Even fewer of the delicate skulls have been retrieved. And until recently we knew nothing about sauropod babies, and we were not even sure that eggs that had been ascribed to sauropods, were, in fact, those of sauropods.

Figure 129. The fossil eggs from Auca Mahuida are the first to be found that contain the embryos of sauropod dinosaurs.

Figure 130. The fossil skin of the embryonic sauropods was preserved in intricate detail. The pebbly texture is not unlike the skin on the leg of a chicken.

In November of 1997, in Argentine's Patagonia region at a location called Auca Mahuida, a team of paleontologists and geologists from the American Museum of Natural History, Yale University and museums in Argentina stumbled across an immense dinosaur nesting ground that had occupied the broad pastoral floodplain over 70 million years ago. Although most eggs were fractured and broken, they found a number of complete eggs measuring about 6 inches across. Paleontologists had long suspected that such large eggs belonged to giant herbivorous sauropods because the fossils of these dinosaurs were found in the same vicinity.

Within some of these fossilized eggs the team discerned small patches of mineralized material with a scaly texture on the surface that looked sus-

piciously like skin. As the search was expanded to the adjacent ridges and ravines, more and larger eggs with this scaly material were found, confirming that what had been discovered was, in fact, the fossilized skin of sauropod embryos. In one ridge, with numerous complete eggs buried just under the surface, several eggs were exposed that contained the tiny bones of embryonic dinosaurs. One of these contained a small hash of minute skull bones and teeth. These miniature, peg-shaped teeth were amazing because they were the same shape as those found in some adult sauropods called titanosaurs. In other eggs, skull bones were discovered shaped like those of titanosaurs. The pattern of scales on some of the skin patches resembled the rose-shaped pattern bones imbedded in the skin of adult titanosaurs, named Saltasaurus. These eggs were most likely laid by a group of these sauropods, which are commonly found in the late Cretaceous sediments of South America.

The eggs are only beginning to be studied, and excavation at the site is continuing. Yet, the eggs at Auca Mahuida are already important for several reasons. First, because the different groups of dinosaurs display characteristic eggshell microstructure, they allow the identification of other sauropod eggs worldwide. Second, they will eventually allow us to tell something about development and changes that occur during growth by comparing the 12 to 14 inch embryos with the nearly 50-foot long adults and with the adults, juveniles and embryos of other dinosaurs.

IV EXPEDITIONS

OF THE AMERICAN MUSEUM OF NATURAL HISTORY

EXPEDITION

The Medicine Bow Anticline

DATE: *1897–1904*

LEADER: *J. I. Wortman and W. Granger*

The Como Bluff region, near the town of Medicine Bow in southeastern Wyoming, has become famous as one of the primary sources for the Jurassic dinosaurs that are featured so prominently in museum exhibits throughout the world. Como Bluff itself is a topographic feature representing part of a folded arch of rock layers called the Medicine Bow Anticline. Dinosaurs were first discovered there in the 1870s by Union Pacific workers building the first transcontinental railroad, the tracks of which were laid near the base of the bluff. News of the rich fossil localities soon reached eastern paleontologists, and by 1877 O. C. Marsh of Yale University had crews of collectors excavating a series of quarries in the region. The Yale expeditions continued for more than a decade, unearthing a wealth of dinosaurs. Workmen hired by the Philadelphia zoologist Edward Drinker Cope, Marsh's rival, had also collected in the region, and this area became one of the flash points of the Marsh-Cope feud of the 1870s and 1880s.

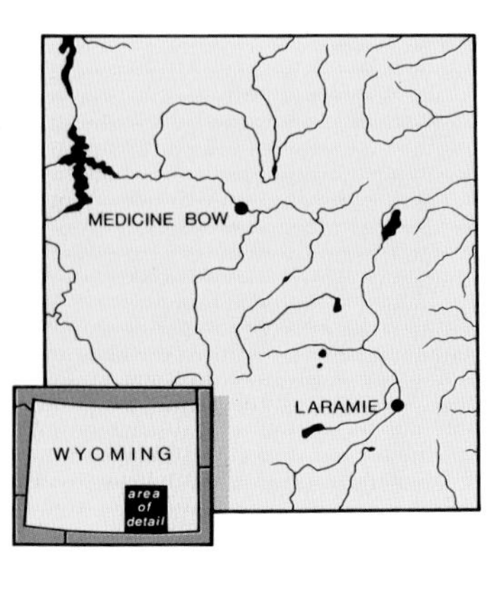

By 1897 Como Bluff itself was nearly depleted of dinosaur fossils. The Museum expeditions, spearheaded by Henry Fairfield Osborn, were not intended to collect dinosaurs; Osborn sent field crews to Como Bluff to search for Jurassic mammals. O. C. Marsh had collected important mammal specimens there 20 years earlier, and much of the Museum's efforts focused on Marsh's old mammal locality, Quarry 9. Work there proved to be fruitless. So in nearby Late Jurassic dinosaur beds the Museum's field parties opened three quarries in 1897. One of these yielded a partial skeleton of giant sauropod *Diplodocus* (AMNH 223), the first dinosaur skeleton excavated by the Museum. Although Osborn found no mammals at Como Bluff, the importance of securing the dinosaur remains for the Musdeum was obvious. His statement in the Museum's annual report of 1897—"*Thus has been inaugurated the second great division fo the work, namely the history of the reptiles in North America*"—paved the way for development of the world's largest dinosaur collection.

FIGURE 131. (SEE PRECEEDING PAGE). **Walter Granger with the first speciman found at Bone Cabin Quarry, a hind limb of the sauropod *Diplodocus* in 1898.**

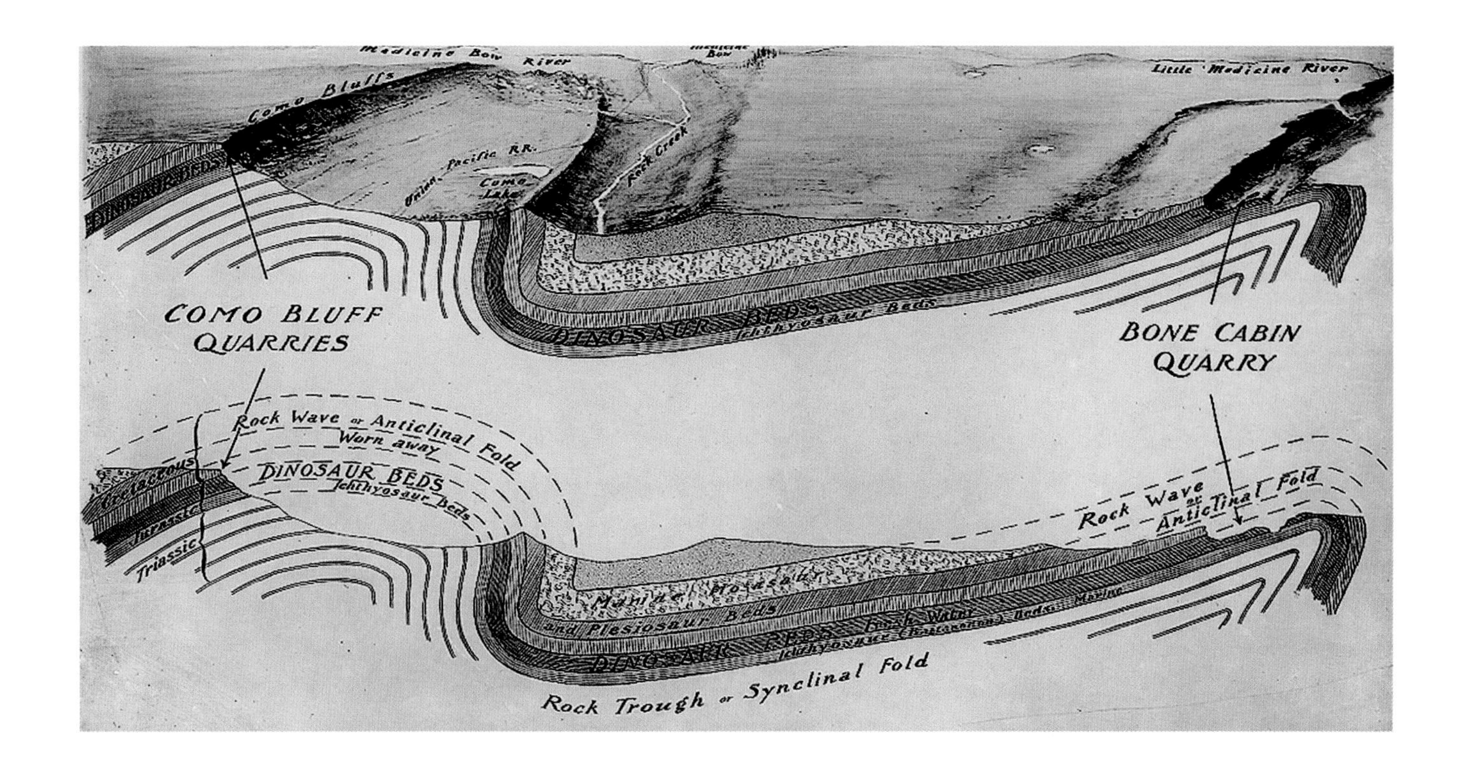

FIGURE 132. The folded strata of the area along the Medicine Bow River in Wyoming has provided more specimens of Jurassic dinosaurs than probably any other location in the world. Here many of the great museums secured some of their prize possessions. The rocks containing these treasures are the Morrison Formation, exposed as a series of ridges called "hog backs" in the local vernacular.

After 1897 Museum field parties began to branch out from the main Como Bluff localities into the adjacent Little Medicine Bow Anticline. During the Jurassic expeditions, the composition of the field parties changed. Barnum Brown left the Late Jurassic effort in 1898 to begin work in South America and the Late Cretaceous beds of the American West. In 1899 Wortman left the Museum to head the newly formed Carnegie Museum's Department of Vertebrate Paleontology. As a result, Walter Granger, who is underappreciated as one of the great fossil collectors and scholars of the field, became leader of the expeditions to the Medicine Bow Anticline.

Beginning in 1898 the Museum's effort centered on Bone Cabin Quarry, named because dinosaur bones were so plentiful there that a local sheepherder had constructed a cabin using the fossils as building stones. Between 1898 and 1903 more than 70,000 kilograms of bones were removed from the site.

Work at Bone Cabin Quarry was strenuous. At least one man overwintered in a small cabin at the site to ward off possible "claim jumpers." Fortunately, the bones were not deeply buried, and overburden was removed using team and scraper (in essence a primitive bulldozer). After specimens

were blocked out, numbered, and their position carefully mapped, they were wrapped in bandages of plaster and burlap and packed in crates with straw padding. Plaster jackets and boxes bore field numbers that were keyed to the quarry map so that their original position could be reconstructed. This excavation marked the first time that these (now standard) methods were used.

The packed crates of dinosaur fossils were carted to the nearby town of Medicine Bow. There they were loaded onto Union Pacific boxcars that, thanks to the influence of J. P. Morgan, Museum trustee, were delivered directly to Manhattan without being reloaded. This direct delivery helped minimize damage to the heavy boxes during transport.

By the early 1900s work at Bone Cabin Quarry had begun to slow down, and the Museum began to develop other prospects. Important among these new sites were Nine Mile Quarry and Quarry R, both in the general vicinity of Bone Cabin Quarry. Although dinosaur bones were not as numerous at these sites, some fine specimens were recovered. Field work in the Medicine Bow Anticline terminated in 1904. Largely through the efforts of Walter Granger, the Museum had acquired tons of bones, including exhibition specimens, such as *Apatosaurus, Ornitholestes, Camptosaurus,* and *Stegosaurus*, that would form the foundation of the dinosaur halls for almost a century.

FIGURE 133. Dinosaur bones from the early expeditions were transported by wagon to the rail head, where they were loaded onto rail cars for the journey to New York. This shipment was facilitated by the railroad magnate J.P. Morgan, an American Museum of Natural History trustee.

EXPEDITION

The Hell Creek Beds

DATE: *1882, 1906*
LEADER: *Barnum Brown*

FIGURE 134. The extensive exposures of the Hell Creek Formation in eastern Montana capture the final chapter of nonavian dinosaur evolution.

The Late Cretaceous beds of the Hell Creek Formation form extensive exposures along the banks of the Missouri River in east-central Montana. These beds preserve one of the best terrestrial Cretaceous-Tertiary boundary sequences in the world, documenting the final chapter in the reign of nonavian dinosaurs (*see* "What was the last dinosaur?," page 61). Hell Creek localities have also produced some of the most spectacular dinosaur fossils in the world.

The formation represents sediments that were deposited by rivers and streams across a broad low floodplain. The floodplain bordered a shallow continental sea that stretched from near the Gulf of Mexico all the way to the Arctic Ocean. The sea acted as a heat sink, ameliorating daily and seasonal climatic extremes. This stable, subtropical climate, along with the abundant supply of water draining off the ancestral Rocky Mountains to the west, supported lush vegetation and a diverse flora and fauna. Not only dinosaurs were present along the floodplain and adjacent areas. Crocodilians, salamanders, turtles, and gars, occupied the rivers, streams, and lakes. Their fossils are preserved in the sands that filled in abandoned stream channels, along with the rarer bones and teeth of small extinct lizards, birds, and mammals. These vertebrates lived alongside a diversity of dinosaurs, including *Tyrannosaurus, Albertosaurus, Triceratops, Anatotitan, Thescelosaurus, Pachycephalosaurus,* and *Edmontosaurus.*

The assemblage of fossils from the Hell Creek Formation provides the best picture of the environment in western North America just before the extinction of the nonavian dinosaurs. This stratigraphic and paleontologic database has served as the basis for most scientific analyses of what caused

One innovation was the construction of a flat-bottomed boat to operate as mobile field headquarters. It measured 4 meters by 10 meters, and was powered by two 7 meter oars, one at each end. The boat was outfitted with a tent that included a stove and small kitchen. From this floating platform the paleontologists worked their way up each side canyon to prospect the surrounding exposures. Following excavation, the packaged specimens were stored on the flatboat. At the end of the season (which often extended well into the freezing weather of late fall), the boat rode low in the water. The weight made the intermittent rapids difficult and dangerous to negotiate. Upon arrival at the main camp, the boxes of bones were packed and shipped while the flatboat was hauled by block and tackle onto the river bank to await the next year's expedition. In later years, a small powerboat supplemented the flatboat. The state of powerboat engineering at the time left something to be desired, as is apparent from Brown's 1912 note to Osborn: *"The motor boat is a success and I am gaining experience as an engineer. It shows more varied moods than a woman, but I still retain a Christian spirit toward it."*

Although in several popular articles Brown exalted the idyllic conditions, photographs of the expedition show that the work was exhausting and had to be conducted in extremes of heat and cold. Often it was so cold that the field party had to wait several days for plaster to set. In images that are almost comical, the group is shown swathed in cheesecloth head coverings to stem the onslaught of vicious northern plains mosquitoes. In some years the water level was low; in others, large amounts of rain submerged the previous year's localities and the river ran with large rapids. In a letter to W. D. Matthew dated July 25, 1915, Barnum Brown described the situation: *"It has rained all over this part of Canada as never before—the Red Deer River is out of banks most of the time; ten feet of water was running over our last year's camp site and came to where our fossils were parked last year. Mosquitoes are fearless of smoke, ferocious and in numbers equal to the Kaiser's army."*

Museum crews were not the only excavators working the Red Deer River beds in the early 1900s. The rich dinosaur beds of Alberta spurred what Edwin Colbert has called the "Canadian Dinosaur Rush." After the much publicized success of Brown's expedition, the Canadian Geological Survey succumbed to local nationalist pressure and contracted the Sternberg family to spearhead its dinosaur collecting expeditions. Since George Sternberg had been a member of the initial Museum expedition, Osborn was perturbed

FIGURE 135. Barnum Brown's 1912 field party, shown protected from mosquitos, ready to set out for prospecting. The steep canyon walls and rough terrain in the Red Deer River area required such novel approaches to field work.

by the newly established competition with the Sternberg family, writing, *"I all too deeply regret that the Ottawa museum which has laid dormant all of these years should suddenly reenter the field."* Matthew concurred that *"it's a pity, but nothing can be done but to beat them to the best specimens."* The two parties were occasional visitors to each others camps, although it can be garnered from Brown's field reports that relations between the groups were strained. At one point, a disgusted Barnum Brown wrote to Matthew that, *"Sternberg and his party are just below us and have taken out some fossils from our territory but we have at present no serious disputes. They have no regard for the ethics of bone digging."*

At the culmination of these expeditions in 1915, the American Museum of Natural History had collected more than three and a half freight cars full of dinosaur bones. The finest specimens (*see "Corythosaurus,"* page 158; *Saurolophus,"* page 153; *Struthiomimus,"* page 121; and *"Albertosaurus,"* page 119), were prepared and put on display immediately. They can be enjoyed now in their newly refurbished environment. Surprisingly, most of these specimens have never been adequately described or illustrated. Other specimens from the Red Deer River, such as the unique skull of *Dromaeosaurus*, are invaluable scientific specimens bearing on a host of contemporary problems and continue to be some of the most heavily studied specimens in our collection.

EXPEDITION

The Flaming Cliffs

DATE: *1922–1925*
LEADER: *Roy Chapman Andrews*

The most famous American Museum of Natural History expeditions were to Central Asia during the early 1920s. These expeditions were the brainchild of Roy Chapman Andrews. Andrews' role in the expeditions and in Museum politics in New York is controversial. To some he was a great expedition leader; to others he was a publicity-seeking adventurer. Most would agree that paleontological research was not his forte. Nevertheless, and to his credit, he spearheaded American Museum of Natural History expeditions in Asia for nearly three decades, recounting his adventures in copious popular works that made him an idol to generations of young, would-be explorers. Andrews had worked in several capacities in the Museum, beginning his career in 1906 as an assistant and janitor in the Department of Mammalogy. From there he worked through the ranks as a technician and a collector. In 1896, while in the employ of the Museum, Andrews entered Columbia University for his Ph.D. study of poorly known Pacific whales. many of his observations were made while he lived in a Japanese whaling village.

FIGURE 136. The Flaming Cliffs get their name from the intense orange color that they take on in the afternoon sun. This photograph was taken in 1992 by Museum paleontologists near the site of the 1925 campsite of the Central Asiatic Expeditions. The area had changed little in the intervening 70 years between Museum expeditions.

The goal of the Central Asiatic Expedition (or, rather, series of expeditions) was billed as an attempt to find the ancestors of modern man. One of Osborn's pet theories was that Central Asia was *"the incubating region for the land life of Europe and North America,"* including human ancestors. Osborn proposed that Andrews lead an expedition there to find them. Andrews outlined an expansive plan to explore the region, collect animals, fossils, and cultural remains, and ship them to the Museum.

These were to be some of the largest and most expensive scientific expeditions ever, requiring over a quarter of a million 1920s dollars. Using his influence in New York society, Osborn paraded Andrews before the country's most influential men, such as J. P. Morgan and John D. Rockefeller. The expedition proceeded in 1922, after more than a year of preparation.

Headquarters was in Beijing, near where Tiananmen Square sits today. The Mongolian segment of the expedition was active during the 1922, 1923, and 1925 field seasons. These trips are chronicled in Andrews' classic narrative with the politically incorrect title *The New Conquest of Central Asia.*

The expeditions were elaborate undertakings, requiring as the first step delicate political negotiations. This was no easy feat, because Mongolia was the second socialist state in the world, declaring its alliance with Moscow in 1922, just four years after the Russian revolution. Mongolia was unstable, with rogue bands of ousted White Russian sympathizers frequently engaging the ruling Bolsheviks. Tribal bandits operating near the Chinese border and the tumultuous political situation in China at the time, made the journey to the field action-packed and punctuated with bursts of gunfire. Although no one was killed during the expeditions, there were many close encounters.

The expedition traveled first to Urga (now Ulaan Baatar) to obtain official permission. Then they proceeded to the southwest, toward the fossil beds. The group traveled in open touring cars, an idea initiated by Andrews, allowing huge distances, up to 160 kilometers, to be traversed in a single day. Along the way the crew met up with camel caravans that had been dispatched to the region several months in advance, carrying fuel supplies and provisions.

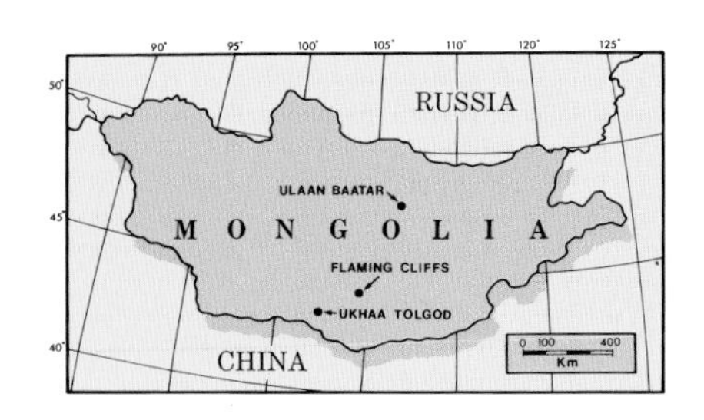

Second-in-command of the expedition, as well as chief paleontologist, was Walter Granger. Granger's role in the expedition was crucial to its success. He directed the paleontologic component of the expedition and found the first fossils of the expedition in 1922 at Iren Dabasu. Granger was also responsible for the fine condition of the specimens, despite their long, arduous trip back to New York, as the following comment by Andrews suggests: *"Granger was pleased at our efforts to discover fossils but his approval ceased abruptly when it came to removing them. My favorite tool was a pick-ax, while he used a camel hair brush and a pointed instrument not much larger than a needle. When a valuable specimen had been discovered he would suggest that we go on a wild ass hunt, or anything that would take us as far as possible from the scene of his operations."*

The discovery of the Flaming Cliffs site has been recounted frequently. According to one version, the field party was traveling to Beijing along an old caravan route. Seeing a yurt in the distance, Andrews stopped the caravan to inquire about local conditions and routes. Granger and the expedition photographer, J. B. Shackelford, went in the other direction to investigate some brightly colored red rocks. When they came to the first exposures, they discovered that the outcrop was just the tip of a large badland. Both Granger and Shackelford knew it was an ideal place to look for fossils.

Within minutes they began to find fossils. Schackelford found a small white skull and a fragmentary egg. Back at the caravan, Granger identified the skull as that of an unknown reptile, and believing that the deposits were Cenozoic, suspected that the eggshell was from a fossil bird. The season was late, and the party could not explore the area before returning to Beijing.

That winter in New York, when the small collection was studied, the importance of the specimens was immediately recognized. The small white skull attracted most of the attention, and was recognized as the remains of a small ceratopsian dinosaur, related to the familiar *Triceratops*.

During the 1923 and 1925 seasons, Museum expeditions collected an impressive array of fossil skeletons at the Flaming Cliffs. These include many small theropods, *Protoceratops*, the ankylosaur *Pinacosaurus*, and a host of lizards and crocodiles. The badlands were called the Flaming Cliffs because of their red-orange color, which often burns brilliantly in the afternoon sun. The locality was called Shabarakh Usu (Mongolian for "muddy spring"), in reference to the well at the base of the cliffs. In wet times this spring often forms a small lake. In the last 10,000 years this lake was permanent, and its banks were common camping sites for nomadic people, whose stone, ceramic, and bronze artifacts are strewn over the dunes. At the the Flaming Cliffs, Museum anthropologists excavated remains of one particularly old culture that they termed the "dune dwellers." Apparently these people also collected dinosaur eggs from the red cliffs, because at their campsites dinosaur eggshells fashioned into beads are found.

Although overshadowed at the time by the discovery of the dinosaur nests, many skeletal specimens collected at the Flaming Cliffs are important to our understanding of contemporary scientific questions, not the least of which is the origin of birds (see *"Saurornithoides,"* page 127; *"Velociraptor,"* page 132; and *"Oviraptor,"* page 124). Aside from the specimens, the times and events associated with these expeditions represent one of the most romantic chapters in the Museum's heritage of exploration.

EXPEDITION

The Western Gobi

DATE: *1991-1995*
LEADER: *Michael J. Novacek*

For more than 60 years after the culmination of the expeditions led by Roy Chapman Andrews (see the previous section), political problems prevented Western scientists from working in Mongolia. The political, economic, and social realignment spurred on in the late 1980s by the end of the Cold War, however, made it possible for American scientists to return to Mongolia. The American Museum of Natural History was approached by the Mongolian Academy of Sciences about the possibility of developing a cooperative field program. In 1990 three members of the Museum's Department of Vertebrate Paleontology (Michael Novacek, Malcolm McKenna, and Mark Norell) traveled to Mongolia to study the feasibility of carrying out expedition work in this legendary fossil field. They negotiated an agreement that called for three full-scale expeditions, to begin in 1991. Unlike the early Museum expeditions, the terms of this agreement stipulated that all the fossils collected remain the property of Mongolia under the auspices of the Mongolian Academy of Sciences. In addition to laying the groundwork for future expeditions, the group made a brief visit to fossil localities in the Gobi Desert. Under the direction of Demberlyin Dashzeveg, the field party left Ulaan Baatar (Mongolia's capital), and two hard-driving, adventure-filled days later they arrived at the Flaming Cliffs. After 66 years Museum paleontologists returned to the spectacular locality discovered by Andrews and Granger in 1922.

FIGURE 137 (ABOVE). The Gobi Desert is one of the great unexplored fossil-bearing regions on Earth. It is only beginning to reveal its treasures as scientists from the American Museum of Natural History, Mongolia, and other countries begin to build on the work of the first explorers.

After visiting the Flaming Cliffs, the party headed south over the Gurvan Saihan Mountains to areas that Andrews and Granger had never reached. Many of these fossil excavation sites in the Nemegt Valley had been discovered by Russian and Polish expeditions; the American Museum of Natural History expeditions would pick up where this work had left off.

In 1992, the first year of the full expedition, work and travel in Mongolia was difficult. A lot has changed since then. The expedition is now in its 11th year. Fossil localities throughout the Southern Gobi have been located, and instead of driving through the desert almost blindly, American Museum paleontologists know the desert better than they know some neighborhoods not far from their homes. Technology has aided also in the form of satellite navigation and imagery. Better vehicles, tastier food, and satellite phones make camp a different place.

In recent years, camp is always made at Ukhaa Tolgad—the site of the most famous discoveries. In ten years this aerially modest locality has supplanted the Flaming Cliffs as not only the richest Mesozoic fossil locality in Mongolia, but the most prolific ever found anywhere. It was here that the nesting dinosaurs were found, the complete skulls of *Shuvuuia* were found,

FIGURE 138 (RIGHT). The embryo of an oviraptorid (below) and the skull of a baby dromaeosaur collected during the 1993 expedition. This is the first embryo of a theropod dinosaur ever found.

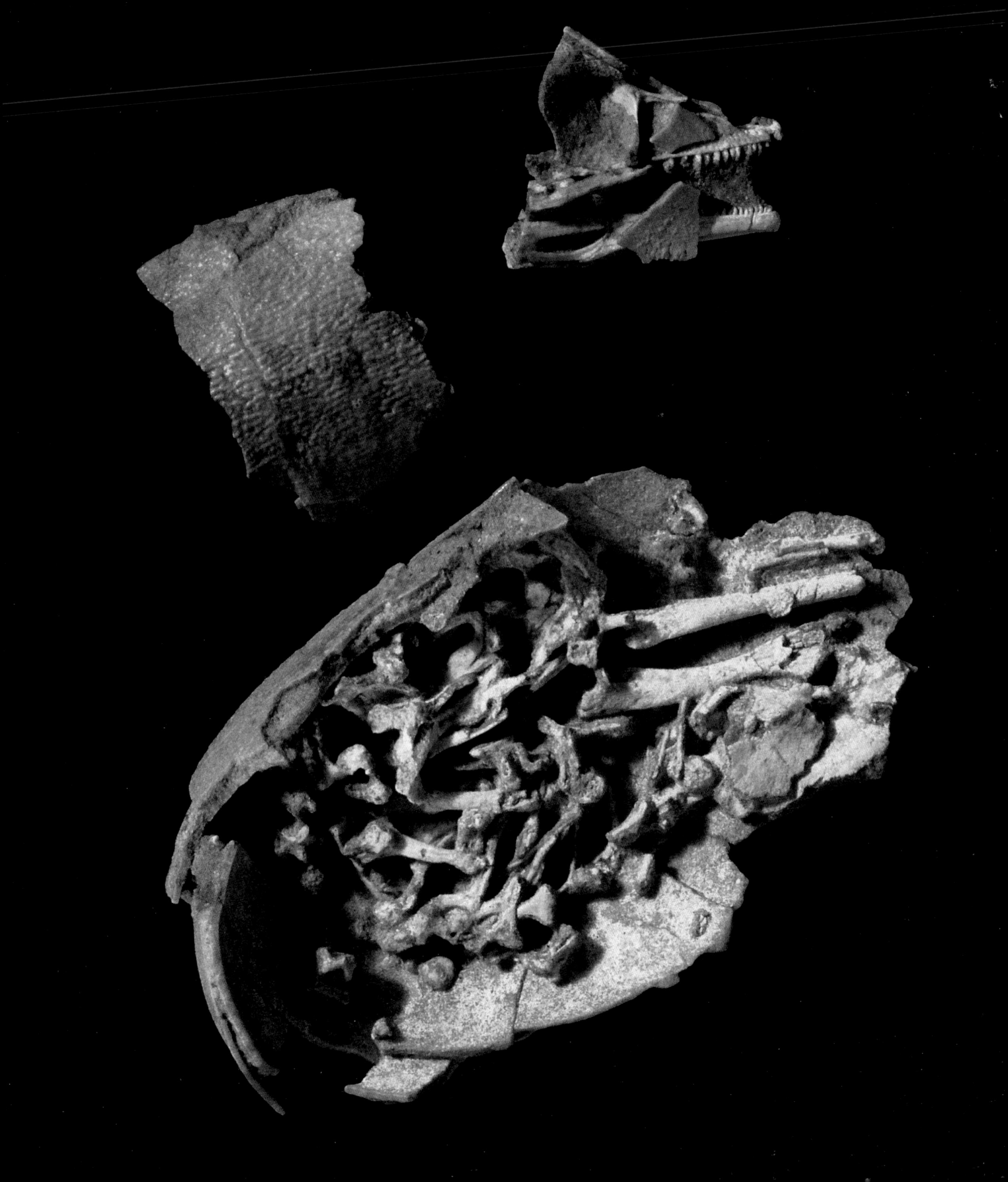

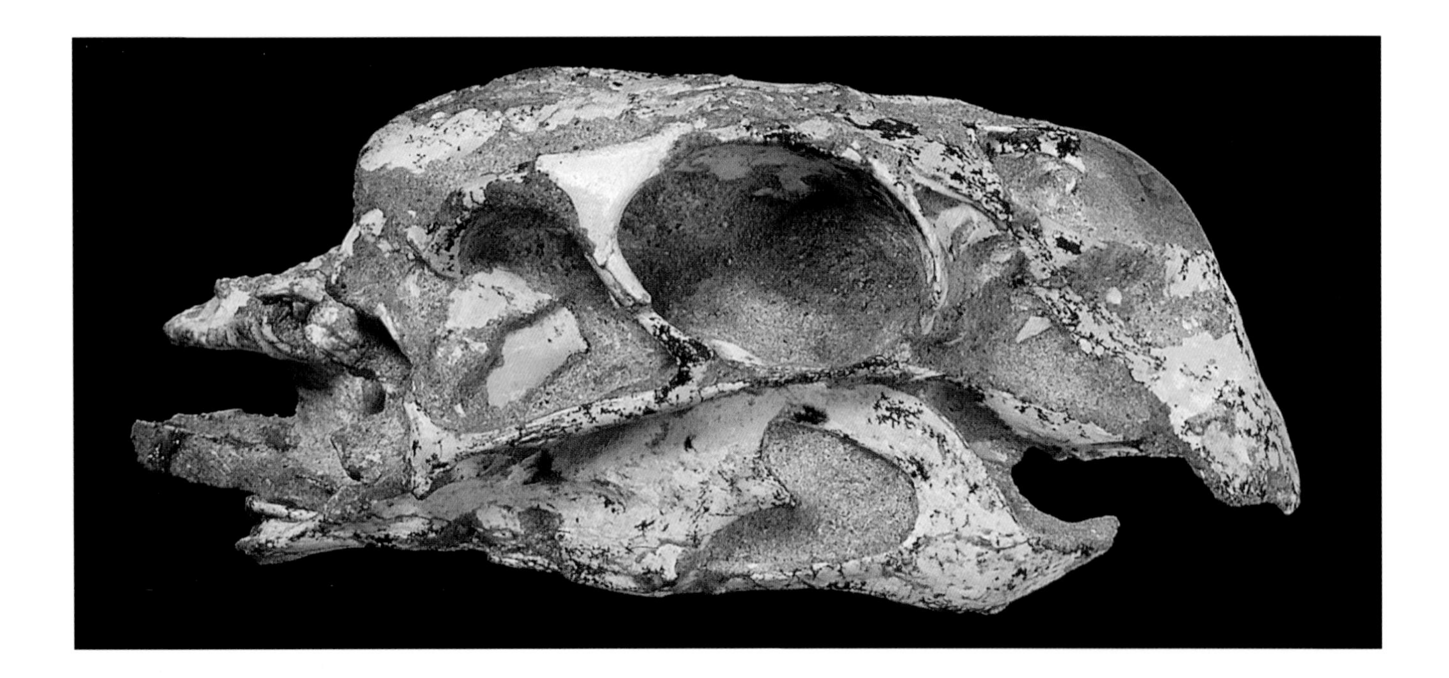

Byronosaurus was found, as well as skeletons of *Pinacosaurus*, protoceratopsians and dromaeosaurs.

As fantastic as Ukhaa Tolgod is for dinosaurs, it is the mammals and lizards that make it really an exceptional site. At this writing, well over 2,000 mammal and lizard fossils have been found. It is rare to find these contemporaries of the dinosaurs, yet here they are abundant. So abundant that more than fl's of all mammal skulls collected during the Cretaceous Period world-wide come from this one small area.

Why are fossils so abundant and well preserved here? It all has to do with the way in which the fossils were preserved. Since fossil specimens were first discovered in the Gobi desert, prevailing theory has been that these animals were covered (some have even proposed smothered) by blowing sand.

Analysis of sediments at Ukhaa Tolgod changed all of that. Here analysis of the geology indicated something different had happened. That instead of blowing sand, it was flowing sand. The sands of Ukhaa Tolgod are peculiar in that the individual sand particles are coated with a layer of clay. This coating of clay makes them essentially slippery. When a sand dune has absorbed enough water, the sand begins to flow like decomposing beach sandcastles. Because the flows were relatively low energy, they did

FIGURE 139. A spectacular specimen of *Ingenia*, a close relative of *Oviraptor*. This specimen was collected at Ukhaa Tolgod during the 1993 expedition. Oviraptorids are very rare. At Ukhaa Tolgod, however, they are one of the most common dinosaurs encountered.

not disturb dead, dying, stubborn or living creatures that much. Consequently, the remarkable preservation. Some of the best-preserved specimens may have been buried in their burrows, as in the case of the mammals and lizards. Others, like the nesters, might have had parental instincts so great that not even an impending sand flow would force them from the nests, a phenomenon that has been observed in living birds south of the High Atlas in Morocco.

One of the most spectacular specimens collected solves a mystery initiated during the 1923 expedition when the first nests of dinosaurs were collected at the Flaming Cliffs. These nests were believed to belong to *Protoceratops*, because bones of this dinosaur are extremely common at that locality. Because no definite associations have been found between *Protoceratops* and the eggs, however, the connection between them has always been conjecture.

A new specimen provides real evidence. This specimen, found in the Nemegt Valley at a locality called Ukhaa Tolgod, is a nest of eggs that were broken open from weathering. Although most of the eggshells were filled with red sand, one egg contained the tiny bones of an embryonic dinosaur. These proved to be the remains of a near-hatchling oviraptorid (*see "Oviraptor,"* page 124). Analysis of the eggshell showed that these eggs were identical to those from the Flaming Cliffs that had been referred to *Protoceratops*.

Associated with one of the original nests, actually lying a few inches on top of it, was the type skeleton of the theropod *Oviraptor* (*see* page 124; see also the previous section). Osborn named this dinosaur *Oviraptor* (literally "egg stealer") because he thought that it had been preying on the eggs of the *Protoceratops*. The name now appears to be somewhat inaccurate, because this animal seems to have died while incubating, caring for, or protecting its nest.

Research on the specimens from Ukhaa Tolgod and on other specimens in the Western Gobi is only beginning. Many spectacular and important specimens have already been collected. Some of these are in the process of preparation. We hope that these specimens will provide us with important information that will render much in this book inaccurate.

EXPEDITION

Liaoning

DATE: 1999–PRESENT

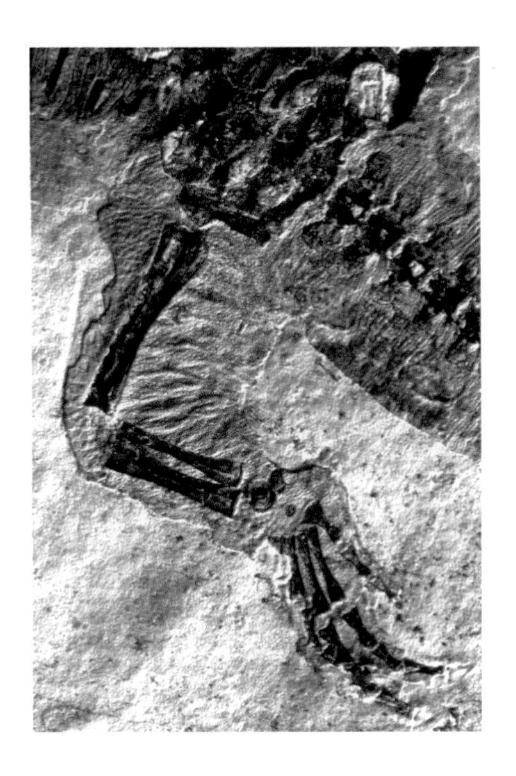

Markets in China are marvelous places. Almost anything imaginable is sold or bartered. Often mixed in with medicinal compounds, plastic dishes, unfamiliar vegetables, and computers are fossils. Early in the 1990's some amazing specimens began to show up in this barely below the surface fossil trade. The remarkable thing about these finds was that they were not just the bones, teeth, or shells—the hard parts of creatures—but impressions of soft parts, including examples of entire animals. The most spectacular fossils among these not only preserved these soft parts, but also often preserved their microstructure and even color pattern. The character of the rock and some of the commoner fossils hinted that these specimens were from Liaoning, an area about 320 kilometers north of Beijing. It wasn't long before Chinese paleontologists were on the trail of what would become one of the most important dinosaur localities ever discovered.

The diversity of fossils at Liaoning is unparalleled and represents one of the most complete community samples recovered from a fossil deposit. A wide array of animals are found here from tiny insects and aquatic arthropods to the remains of sizable dinosaurs. Fish with stripes, a turtle with spots, mammal relatives covered with hair, pterosaurs with leathery wings, and

blooming plants have been found. You have read about some of the Liaoning feathered dinosaurs (*see* "*Confuciusornis,*" page 186, and "*Caudipteryx*," page 188). Others include *Sinosauropteryx*, a small animal covered with what may be rudimentary feathers, a therizinosaur (*Beipaosaurus*) with feather-like structures that line the back of its arms, and even reports of a dromaeosaur that is covered with a feathery body covering.

FIGURE 140. (OPPOSITE) Many fossils from Liaoning are so finely preserved that they even show the textures of skin as on this leg of a champsosaur, a small lizard-like aquatic reptile.

FIGURE 141. Plant fossils are extremely well preserved at Liaoning, they include remarkable specimens like this gymnosperm inflorescence.

This fine preservation is a result of how the specimens were originally deposited. When these animals lived, the Liaoning area was dotted with large, still lakes. Intermittent regional volcanic eruptions filled the skies with clouds of fine volcanic ash. As the ash settled, much of it was deposited in lakes—quickly covering dead animals lying on the bottom with layers of extremely fine sediment. Because the sediments are so fine and the animal and plant remains were not disturbed by the action of running water or the depredations of scavengers, the carcasses were protected for millions of years, preserving the finest details of skeleton and viscera. Although the exact chemical reason for the preservation of the soft tissue is only now beginning to be studied, it is likely that the volcanic ash caused a microenvironment, perhaps depleted in oxygen, modifying the typical actions of decomposing bacteria.

The exact age of the Liaoning deposits is disputed. Many of the animals, like long-tailed pterosaurs and some of the primitive mammal relatives, are similar to creatures that lived in Europe about 140 million years ago—roughly at the same time as the Late Jurassic proto-bird *Archaeopteryx*. Others, like many of the fishes, look more like Early Cretaceous species. Recently using radioactive dating (*see* 81), a team of Chinese and American scientists determined that at least some of the beds were 125 mil-

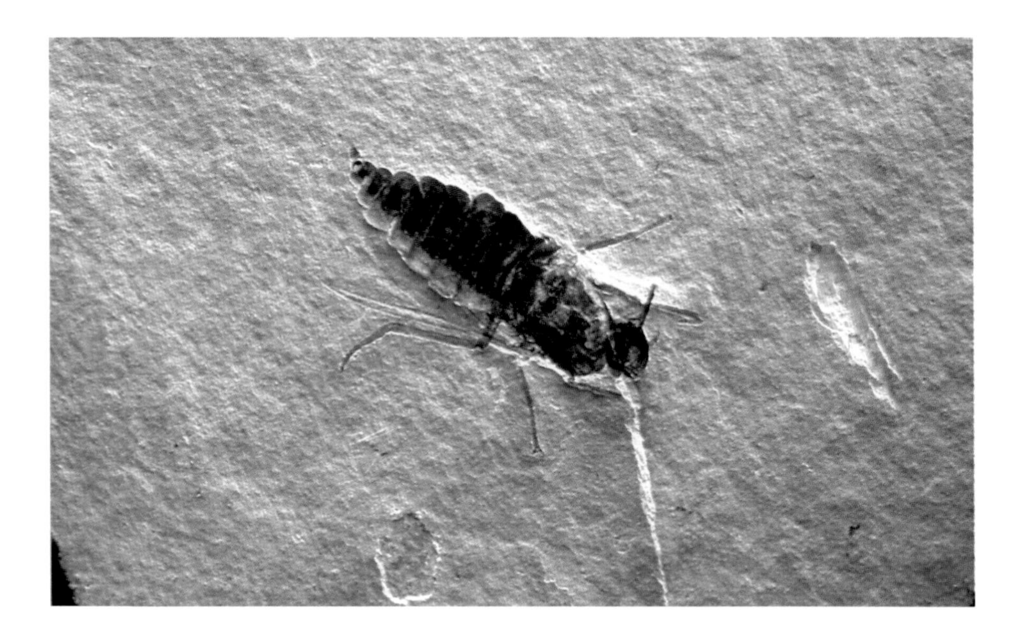

lion years old, or Early Cretaceous in age. The Liaoning deposits however, are extensive and complex, so this date pertains only to the site the samples were taken. Other places in the same area that have fossils may be older or younger.

FIGURE 142. The sediments of Liaoning are important because they preserve all elements of the biota—from dinosaurs and mammals to plants and insects. Here a small insect even shows a banded color pattern.

Collecting at Liaoning is not your usual fossil hunting pursuit. Most of the specimens have been dug out of the ground by local farmers hoping to supplement their meager incomes with the discovery of a great fossil treasure. Although illegal, huge quantities of fossil remains have, and continue to be, removed in this way—so much so that illicit quarrying operations dot the hillsides. Most of these specimens find their way into the international curio market after they have been smuggled out of China. Obviously these specimens are both lost to science, and when available to scientists cannot be incorporated into reputable museum collections whose policies prohibit acquisition of illegally collected materials.

Liaoning is one of the most exciting places of fossil discovery on the planet. Currently major excavations, led by several groups, including the American Museum cooperating with Chinese scientists, have revealed never before seen animals and plants. The paleontological riches of this place are unimaginable. Not even the initial chapter on the Liaoning fossils has been written, and we have learned so much already. The future course of dinosaur research is certain to be significantly changed by the discovery of more unique specimens at this fertile location.

INDEX

INDEX

INDEX

ILLUSTRATION CREDITS

Department of Vertebrate Paleontology Archives, American Museum of Natural History: (figure number) 2, 3, 4, 5, 6, 7, 9, 10, 11, 14, 17, 18, 21, 22, 23, 24, 26, 27, 30, 31, 34, 36, 40, 41, 43, 46, 52, 55, 56, 59, 60, 61, 63, 64, 66, 67, 69, 72, 73, 74, 76, 77, 78, 79, 80, 81, 82, 84, 88, 91, 92, 94, 96, 97, 98, 99, 100, 101, 102, 103, 104, 105, 108, 110, 111, 112, 113, 114, 115, 117, 118, 119, 120, 122, 123, 124, and 126. Library Photography Collection, American Museum of Natural History: 1, 25, 35, and 106. Michael Ellison, Department of Vertebrtate Paleontology, American Museum of Natural History: 8, 15, 20, 28, 29, 32, 33, 38, 39, 45, 50, 65, 70, 83, 85, 86, 87, 89, 90, 93, 95, 109, 116, 121, 129, and 130. David Grimaldi, Department of Entomology, American Museum of Natural History: 48. Lowell Dingus, Project Director, Fossil Halls Renovation, American Museum of Natural History: 44 and 125. Fred Conrad: 13 and 128. Mark A. Norell, Department of Vertebrate Paleontology, American Museum of Natural History: 16, 37, 49, 51, 53, 54, 57, 58, and 127. Brian Kosoff: 12 and 107. Murray Alcosser: 42. Robert A. Tyrrell: 19. Science Photo Library: 47. Photography Studio, American Museum of Natural History: 62. Ben Blackwell/Craig Chesek/Denis Finnin, © 1995 American Museum of Natural History: 68, 71, and 75.

ACKNOWLEDGMENTS

We would like to thank our colleagues in the Department of Vertebrate Paleontology, American Museum of Natural History, and the staff of the Fossil Halls Renovation. The work of these dedicated men and women is a continuation of a tradition of excellence in paleontology research and display at the American Museum of Natural History. The American Museum of Natural History Board of Trustees is recognized for their support of the Fossil Halls Renovation and related endeavors. Several individuals around the Museum greatly contributed to the development of this book. Among these we would especially like to thank Robert Gebbie, Jeanne Collins, Denis Finnin, Craig Chesek, Jackie Beckett, David Grimaldi, Scarlett Lovell, Anna Schermerhorn, Jim Clark, Mike Novacek, and Joel Sweimler. Vivian Pan worked hard as a proofreader and photography editor throughout the project. Much of the artistic direction of this book was set by the photographs and excellent counsel of Michael Ellison. At Alfred A. Knopf Susan Ralston, Amanda Gordon, and Susan Carroll were early and enthusiastic supporters of the book, as was Vivien Bowler at Little, Brown and Company (U.K.). Barbara Balch is credited for her excellent book design. Brian Kosoff, Ed Heck, Robert A. Tyrrell, and Murray Alcosser are thanked for providing images, and Nicole Potter for the index. Henry Galiano, of Maxilla and Mandible, Donna David, and Ann J. Perrini, President of Nevraumont Publishing Company, all made valuable contributions. Finally, William Moynihan stands out as both a supporter of this project, the Fossil Halls Renovation, and the Museum as a whole. Without his administrative talents and hard work, this book, the beautiful new dinosaur halls, and other projects would never have seen fruition.